Advanced Circuit I

Electroni

Aleksander Petrov

Author: Aleksander Petrov

Title: Advanced Circuit Design Techniques for

This book is a product of [Publisher's Aleksander Petrov]

ISBN:

TABLE OF CONTENTS

Chapter 1: Introduction to Advanced Circuit Design Techniques

Overview of Circuit Design Principles

Introduction:

In the rapidly evolving field of electronics engineering, circuit design principles play a pivotal role in the development of cutting-edge electronic devices and systems. This subchapter aims to provide engineers, especially electrical engineers, with a comprehensive overview of the fundamental principles that underpin the design of electronic circuits. Understanding these principles is essential for engineers to create efficient, reliable, and innovative electronic solutions.

Basic Concepts:

The subchapter begins by introducing the fundamental concepts of electronic circuits, such as voltage, current, resistance, and power. These concepts serve as the building blocks for more complex circuit designs and are crucial for engineers to grasp. Additionally, the concept of Ohm's Law is explained, illustrating the relationship between voltage, current, and resistance.

Analog and Digital Circuits:

This section distinguishes between analog and digital circuits, highlighting their unique characteristics and design considerations. Analog circuits process continuous signals, while digital circuits operate on discrete signals. The subchapter explores the various components and techniques used in both analog and digital circuit design, including operational amplifiers, transistors, logic gates, flip-flops, and registers.

Circuit Analysis Techniques:

The subchapter delves into different circuit analysis techniques necessary for engineers to analyze and evaluate circuit performance. It covers essential methods, such as Kirchhoff's laws, mesh analysis, nodal analysis, Thevenin's theorem, and superposition theorem. These techniques enable engineers to understand circuit behavior, calculate voltage and current values, and optimize circuit performance.

Circuit Design Considerations:

This section emphasizes the importance of considering several factors during the circuit design process. It explores topics such as power consumption, noise immunity, signal integrity, thermal management, and electromagnetic compatibility. Engineers learn how to select appropriate components, make trade-offs, and ensure the reliability and efficiency of their circuit designs.

Advanced Circuit Design Techniques:

The final part of the subchapter provides an overview of advanced circuit design techniques employed in modern electronic systems. Topics covered include frequency response analysis, feedback systems, filter design, impedance matching, signal conditioning, and mixed-signal circuit design. These advanced techniques equip engineers with the skills to tackle complex challenges encountered in real-world circuit design applications.

Conclusion:

The subchapter on the overview of circuit design principles serves as a foundation for engineers, particularly electrical engineers, to navigate the intricate realm of electronic circuit design. By comprehending the basic concepts, analysis techniques, design considerations, and advanced techniques discussed, engineers can confidently embark on designing innovative, reliable, and high-performance electronic circuits for a wide range of applications.

Importance of Advanced Circuit Design Techniques

In the ever-evolving field of electrical engineering, staying ahead of the game is crucial for engineers to design cutting-edge electronic devices. Advanced circuit design techniques play a pivotal role in achieving this goal. This subchapter aims to highlight the importance of these techniques for electrical engineers and explore how they can enhance the overall performance, efficiency, and reliability of electronic circuits.

One of the primary reasons advanced circuit design techniques are crucial is their ability to optimize the performance of electronic devices. By employing advanced algorithms and methodologies, engineers can fine-tune circuit parameters and achieve higher levels of functionality. This results in improved signal integrity, reduced noise interference, and enhanced overall system performance. Whether it is designing high-speed communication systems, power electronics, or low-power devices, advanced circuit design techniques enable engineers to push the boundaries of what is possible.

Moreover, these techniques play a vital role in enhancing the efficiency of electronic circuits. With the increasing demand for energy-efficient devices, engineers must focus on minimizing power consumption while maximizing performance. Advanced circuit design techniques, such as low-power design methodologies, enable engineers to achieve this delicate balance. By leveraging innovative approaches like dynamic voltage and frequency scaling, power gating, and clock gating, engineers can design circuits that consume minimal power without compromising on functionality.

Reliability is another critical aspect that advanced circuit design techniques address. As electronic devices become more complex and miniaturized, the risk of failures and malfunctions increases. Advanced techniques like fault detection and self-repair mechanisms allow engineers to design circuits that are more robust and resilient to failures. By implementing redundancy, error correction codes, and fault-tolerant designs, engineers can ensure the reliability and longevity of electronic systems.

Lastly, advanced circuit design techniques foster innovation and enable engineers to explore new frontiers. These techniques encourage engineers to think outside the box, experiment with novel circuit topologies, and utilize emerging technologies. By pushing the boundaries of conventional design methods, engineers can unlock new possibilities and create breakthrough solutions in diverse fields such as Internet of Things (IoT), wearable technology, and renewable energy.

In conclusion, advanced circuit design techniques are of utmost importance for electrical engineers. They enable engineers to optimize performance, enhance efficiency, ensure reliability, and foster innovation. By incorporating these techniques into their design process, engineers can stay at the forefront of technological advancements and create electronic devices that revolutionize the way we live and interact with technology.

Challenges in Modern Circuit Design

In the rapidly evolving field of electronics, electrical engineers face numerous challenges when designing circuits for modern applications. With the constant demand for smaller, faster, and more efficient devices, engineers must navigate through a myriad of obstacles to create cutting-edge circuit designs. This subchapter explores the key challenges that engineers encounter in modern circuit design and provides valuable insights into overcoming them.

One of the primary challenges in modern circuit design is miniaturization. As technology advances, there is an increasing need for smaller and more compact electronic devices. However, cramming complex circuitry into a smaller space poses several difficulties. Engineers must find innovative ways to reduce component size, optimize power consumption, and minimize heat dissipation while ensuring optimal performance.

Another significant challenge is the ever-increasing demand for speed and performance. As data processing requirements continue to rise, engineers must design circuits that can handle high-speed data transfer and complex computations. This necessitates the use of advanced techniques such as signal integrity analysis, noise reduction, and high-frequency design to ensure reliable operation and minimize signal degradation.

Power management is also a critical challenge in modern circuit design. With the proliferation of portable devices and the growing concern for energy efficiency, engineers must develop circuits that consume minimal power while delivering optimal performance. This involves employing techniques like low-power design, power gating, and efficient voltage regulation to maximize battery life and reduce overall power consumption.

Furthermore, the integration of various technologies poses a significant challenge for engineers. As circuit designs become more complex, combining different technologies such as analog, digital, and RF circuits requires careful consideration. Engineers must overcome compatibility issues, interference problems, and ensure seamless integration of different components to achieve the desired functionality.

Lastly, the ever-changing nature of technology demands continuous learning and adaptation. Engineers must stay updated with the latest advancements and emerging trends in circuit design. They need to embrace new tools, methodologies, and technologies to keep pace with the rapid evolution of the field.

In conclusion, the challenges in modern circuit design require electrical engineers to be innovative and adaptable. From miniaturization and speed to power management and integration, engineers must address these challenges to create state-of-the-art circuit designs. By staying abreast of the latest advancements and employing advanced techniques, engineers can overcome these obstacles and contribute to the development of cutting-edge electronic devices.

Chapter 2: Signal Integrity and Noise Analysis

Understanding Signal Integrity Issues

In the ever-evolving field of electronics engineering, it is crucial for electrical engineers to have a deep understanding of signal integrity issues. As electronic devices become more complex and high-speed, the challenges related to maintaining signal integrity have become increasingly paramount. This subchapter aims to provide engineers with a comprehensive overview of signal integrity issues and equip them with the necessary knowledge and techniques to address these challenges effectively.

Signal integrity refers to the quality of electrical signals as they propagate through various components and interconnects in a circuit. Maintaining signal integrity is crucial for reliable and efficient operation of electronic systems. Signal integrity issues can arise due to a variety of factors such as noise, distortion, crosstalk, reflections, and impedance mismatches. Failure to address these issues can lead to data errors, reduced performance, and even system failure.

This subchapter will delve into the fundamentals of signal integrity, starting with an explanation of the key parameters that affect signal quality, such as rise time, jitter, and eye diagram analysis. It will explore the various sources of signal degradation, including transmission line effects, power integrity issues, electromagnetic interference (EMI), and grounding problems. The subchapter will also discuss the impact of signal integrity issues on high-speed digital and analog circuits, as well as wireless communication systems.

To mitigate signal integrity issues, engineers must employ a variety of design techniques and tools. This subchapter will provide an overview of common design practices, including proper PCB layout, transmission line termination, decoupling capacitor placement, and controlled impedance routing. It will also introduce advanced simulation and analysis tools that aid in identifying and resolving signal integrity issues during the design stage.

Moreover, the subchapter will highlight real-world examples and case studies, demonstrating the consequences of overlooking signal integrity issues and showcasing successful solutions implemented by experienced engineers. It will encourage engineers to adopt a proactive approach to signal integrity, emphasizing the importance of thorough testing, validation, and post-layout analysis.

By thoroughly understanding and addressing signal integrity issues, electrical engineers can ensure the reliable and optimal performance of electronic systems. As technology continues to advance at a rapid pace, mastering signal integrity principles will be indispensable for engineers seeking to design cutting-edge circuits and meet the ever-increasing demands of today's electronic applications.

Noise Sources in Electronic Circuits

Noise is an inevitable phenomenon in electronic circuits that can significantly impact the performance and reliability of electronic devices. As electrical engineers, it is crucial to understand the various noise sources that can affect circuit operation and take appropriate measures to mitigate their effects. This subchapter titled "Noise Sources in Electronic Circuits" aims to provide engineers with a comprehensive overview of the different noise sources encountered in electronic circuits and their implications.

The subchapter begins by introducing the concept of noise and its significance in electronic circuits. It explains how noise can arise from both external and internal sources, including thermal noise, shot noise, flicker noise, and interference from other electronic devices. Each noise source is described in detail, highlighting its characteristics, statistical properties, and mathematical models.

Furthermore, the subchapter delves into the impact of noise on circuit performance. It discusses how noise can degrade signal quality, reduce the signal-to-noise ratio, and limit the dynamic range of a circuit. Engineers are provided with practical examples that illustrate the detrimental effects of noise on various electronic devices, such as amplifiers, filters, and communication systems.

To address these challenges, the subchapter explores different techniques and strategies to minimize noise in electronic circuits. It discusses the importance of proper grounding and shielding techniques, component selection, and circuit layout optimization. Moreover, it covers advanced noise reduction techniques, including feedback systems, noise-canceling circuits, and low-noise design principles.

To enhance the understanding of engineers, the subchapter includes numerous illustrations, diagrams, and simulation results that demonstrate the concepts discussed. Real-world case studies and practical examples are also provided to illustrate the application of noise reduction techniques in different electronic circuits.

In conclusion, "Noise Sources in Electronic Circuits" is a crucial subchapter within the book "Advanced Circuit Design Techniques for Electronics Engineers." It equips electrical engineers with a comprehensive understanding of noise sources in electronic circuits and empowers them with the knowledge to effectively mitigate noise and optimize circuit performance. By mastering these concepts, engineers will be able to design robust and reliable electronic systems that meet the stringent noise requirements of modern applications.

Techniques for Signal Integrity Analysis

In the rapidly evolving field of electronics, ensuring proper signal integrity is crucial for the successful design and implementation of complex circuits. Signal integrity refers to the ability of a signal to propagate through a circuit without distortion or degradation. With the increasing complexity and speed of modern electronic systems, engineers, especially electrical engineers, need to be equipped with advanced techniques for signal integrity analysis.

This subchapter explores various techniques that electrical engineers can employ to analyze and improve signal integrity in their designs.

Firstly, it delves into the importance of high-frequency design considerations. It explains the challenges that arise when dealing with high-speed signals, such as impedance matching, crosstalk, and electromagnetic interference (EMI). The subchapter then introduces techniques like termination schemes, transmission line design, and decoupling capacitors that can mitigate these issues and enhance signal integrity.

Next, it discusses the significance of simulation and modeling in signal integrity analysis. It highlights the use of advanced software tools, such as SPICE (Simulation Program with Integrated Circuit Emphasis) and electromagnetic simulation software, to accurately predict signal behavior and identify potential issues. The subchapter provides practical tips on how to set up simulations and interpret the results to optimize signal integrity.

Furthermore, the subchapter explores measurement techniques for signal integrity analysis. It covers the use of tools like oscilloscopes, vector network analyzers, and time domain reflectometers to measure signal quality parameters like rise time, overshoot, and jitter. It also explains how engineers can utilize these measurements to troubleshoot and fine-tune their circuit designs.

Additionally, the subchapter addresses the growing importance of signal integrity analysis in high-speed printed circuit board (PCB) designs. It introduces techniques like controlled impedance routing, layer stackup design, and power distribution network optimization to minimize signal degradation and ensure reliable signal transmission.

Lastly, the subchapter concludes with a discussion on the future trends and challenges in signal integrity analysis. It touches upon emerging technologies like high-speed serial interfaces, system-level simulation, and the impact of miniaturization on signal integrity.

Overall, this subchapter on techniques for signal integrity analysis equips electrical engineers with the essential knowledge and tools to tackle the complex challenges associated with ensuring proper signal propagation in modern electronic circuits. By implementing these techniques, engineers can enhance the performance, reliability, and efficiency of their designs, thereby advancing the field of electronics.

Chapter 3: High-Speed Digital Circuit Design

Introduction to High-Speed Digital Circuits

In today's rapidly advancing technological landscape, high-speed digital circuits have become the backbone of countless electronic devices, ranging from smartphones and tablets to advanced medical equipment and autonomous vehicles. As an electrical engineer, it is crucial to have a solid understanding of the principles and techniques involved in designing and implementing these circuits.

This subchapter aims to provide you, as an engineer, with a comprehensive introduction to high-speed digital circuits. We will explore the fundamental concepts, challenges, and design considerations involved in the development of these circuits.

To begin, we will delve into the underlying principles of digital circuits, including binary logic, Boolean algebra, and basic gate-level designs. Understanding these foundational concepts is essential for tackling more complex high-speed circuitry.

Next, we will explore the unique challenges that arise when working with high-speed digital circuits. These challenges include signal integrity issues, noise, crosstalk, and power consumption. We will discuss various techniques to mitigate these challenges, such as proper grounding and shielding, impedance matching, and advanced signal routing strategies.

Furthermore, this subchapter will provide an overview of the various types of high-speed digital circuits commonly used in modern electronics. We will discuss the characteristics and applications of flip-flops, registers, counters, and memory elements, among others. Additionally, we will explore the concept of clock distribution and synchronization, which is critical for maintaining the integrity of digital signals in complex systems.

Moreover, this subchapter will introduce you to the different design methodologies and tools available for high-speed digital circuit design. We will explore the use of hardware description languages (HDLs), simulation and modeling techniques, and the importance of verification and testing in ensuring the reliability and performance of these circuits.

Finally, we will conclude this subchapter by providing an overview of recent advancements and emerging trends in high-speed digital circuit design. Topics such as high-speed serial interfaces, clock and data recovery, and low-power design techniques will be explored, giving you a glimpse into the future of this rapidly evolving field.

By the end of this subchapter, you will have gained a solid foundation in high-speed digital circuit design. Armed with this knowledge, you will be well-equipped to tackle the challenges and complexities associated with designing cutting-edge electronics in today's fast-paced technological landscape.

Whether you are a seasoned electrical engineer or a novice in the field, this subchapter will serve as an invaluable resource to enhance your understanding of high-speed digital circuits and enable you to excel in your role as a modern electronics engineer.

Transmission Line Theory for High-Speed Signals

In the digital age, high-speed signals have become ubiquitous in various applications including telecommunications, data centers, and consumer electronics. As a result, engineers need to have a solid understanding of transmission line theory to effectively design and analyze circuits that can handle these high-frequency signals with minimal distortion and signal degradation.

Transmission line theory provides a framework for understanding the behavior of electrical signals as they propagate along conductors over long distances or through various media. It takes into account the distributed nature of the transmission line, accounting for effects such as signal reflection, attenuation, and dispersion.

One of the key concepts in transmission line theory is the characteristic impedance, which determines how a transmission line will react to signals of different frequencies. By matching the characteristic impedance of the transmission line with the source and load impedances, engineers can minimize signal reflections and maximize power transfer efficiency.

Another important aspect of transmission line theory is signal integrity. High-speed signals are prone to various distortions and noise, such as ringing, crosstalk, and electromagnetic interference. Understanding transmission line theory allows engineers to design circuits with proper termination, impedance matching, and routing techniques to minimize these unwanted effects and ensure reliable signal transmission.

In this subchapter, we will delve into the fundamental principles of transmission line theory for high-speed signals. We will discuss the basics of transmission line models, including the lumped-element model and the distributed-element model, highlighting their strengths and limitations. We will also explore the different types of transmission lines commonly used in high-speed circuit design, such as microstrip, stripline, and coaxial cables, and discuss their respective advantages and trade-offs.

Furthermore, this subchapter will cover advanced topics in transmission line theory, including signal propagation, signal integrity analysis, and the impact of transmission line discontinuities on signal quality. We will also introduce simulation tools and techniques for analyzing transmission line behavior, such as SPICE models and electromagnetic field solvers.

By the end of this subchapter, engineers specializing in electrical engineering will have a comprehensive understanding of transmission line theory for high-speed signals. They will be equipped with the knowledge and tools necessary to design robust and efficient circuits that can handle the challenges posed by high-frequency signals in today's demanding applications.

Design Considerations for High-Speed PCB Layout

In today's fast-paced world of electronics, high-speed printed circuit board (PCB) design has become a critical aspect for electrical engineers. As technology advances and data transfer rates continue to rise, it is essential to understand the design considerations necessary for high-speed PCB layout. This subchapter aims to provide engineers with valuable insights and guidelines to successfully tackle these challenges.

First and foremost, signal integrity is of utmost importance when dealing with high-speed PCB layout. Engineers must carefully consider the transmission lines' impedance matching, termination techniques, and crosstalk mitigation. These factors play a crucial role in ensuring minimal signal degradation and maintaining high data transfer rates.

To achieve optimal signal integrity, engineers should pay close attention to the routing techniques employed in the PCB design. Differential pair routing, which involves routing two signals with equal and opposite phases, is commonly used to reduce noise and enhance signal quality. Additionally, the use of controlled impedance traces and proper layer stacking techniques can minimize signal reflections and improve overall performance.

Another crucial consideration is power integrity. High-speed designs often require multiple power domains, and it is essential to carefully plan and implement power distribution networks (PDNs). Engineers should analyze the power requirements of each component, optimize power delivery, and minimize voltage fluctuations and noise. Proper decoupling capacitor placement and selection are also vital to maintain stable power supplies and mitigate switching noise.

Thermal management is another critical aspect of high-speed PCB layout. With increased data rates, components tend to generate more heat. Engineers must ensure efficient heat dissipation by considering proper component placement, utilizing heat sinks, and providing adequate airflow. Thermal analysis and simulation tools can assist in identifying potential hotspots and optimizing the design for optimal cooling.

Furthermore, the selection of high-quality materials and components is crucial for high-speed PCB layout. Engineers should carefully choose PCB substrates with low dielectric loss and consistent characteristics throughout the board. High-speed connectors, transmission lines, and components with low insertion loss and impedance variation should be prioritized to ensure reliable signal transmission.

In conclusion, high-speed PCB layout requires careful consideration of several design aspects. Signal integrity, power integrity, thermal management, and component selection are all crucial considerations for successful high-speed designs. By adhering to these guidelines, electrical engineers can achieve optimal performance, reduce signal degradation, and ensure the reliability of their high-speed PCB designs.

Chapter 4: Analog Circuit Design Techniques

Basics of Analog Circuit Design

Analog circuit design forms the foundation of modern electrical engineering, enabling engineers to create circuits that process, amplify, and manipulate continuous signals. Understanding the basics of analog circuit design is crucial for electrical engineers as they strive to develop innovative solutions for a wide range of applications. In this subchapter, we will delve into the fundamental principles and techniques that underpin analog circuit design.

The first step in analog circuit design is gaining a solid understanding of basic electronic components such as resistors, capacitors, and inductors. These passive components play a vital role in shaping the behavior of analog circuits. We will explore their characteristics, properties, and the mathematical models used to describe their behavior in various circuit configurations.

Next, we will examine different types of amplifiers, including operational amplifiers (op-amps), which are widely used in analog circuit design. Op-amps are versatile devices that offer high gain and excellent performance, making them indispensable in a wide range of applications. We will discuss various op-amp configurations, such as inverting and non-inverting amplifiers, and explore techniques to optimize their performance.

Furthermore, we will explore the concept of feedback in analog circuits. Feedback is a crucial design principle that affects stability, bandwidth, and overall performance. We will discuss the different types of feedback, including positive and negative feedback, and their impact on circuit behavior. Additionally, we will investigate techniques such as compensation and stabilization to overcome potential instability issues.

In addition to amplifiers, filters are another essential component of analog circuit design. Filters allow engineers to manipulate the frequency content of signals, enabling applications like signal conditioning, noise reduction, and frequency response shaping. We will explore different types of filters, including passive filters and active filters, and discuss their design considerations, trade-offs, and practical implementation.

Finally, we will touch upon the importance of noise analysis and mitigation in analog circuit design. Noise can degrade the performance of analog circuits, affecting signal integrity and accuracy. We will discuss various sources of noise, such as thermal noise, shot noise, and flicker noise, and explore techniques to minimize their impact on circuit performance.

By mastering the basics of analog circuit design, engineers will be equipped with the necessary knowledge and skills to design, analyze, and optimize analog circuits for a variety of applications. This subchapter serves as a stepping stone for engineers to delve deeper into advanced circuit design techniques, enabling them to create innovative and robust solutions in the field of electrical engineering.

Operational Amplifiers and Their Applications

In the fascinating world of electronics engineering, operational amplifiers (op-amps) have emerged as a fundamental building block for a wide array of applications. These versatile devices, with their ability to amplify and manipulate electrical signals, have revolutionized the field and propelled technological advancements in various industries. This subchapter delves into the intricacies of operational amplifiers and explores their diverse applications.

The subchapter begins by providing engineers, particularly electrical engineers, with a comprehensive understanding of op-amps. It presents a detailed overview of their internal structure, including the inputs, outputs, and power supply connections. By exploring the inner workings of op-amps, engineers gain valuable insights into their characteristics, such as high gain, high input impedance, and low output impedance. This knowledge serves as a foundation for the subsequent discussions on their applications.

One of the primary applications covered in this subchapter is signal amplification. Op-amps play a pivotal role in amplifying weak signals to usable levels, enabling engineers to extract meaningful information from sensors, transducers, and other low-level signal sources. The subchapter explores various amplifier configurations, including inverting amplifiers, non-inverting amplifiers, and differential amplifiers, elucidating their design principles and performance trade-offs.

Another important aspect addressed in this subchapter is op-amp-based filters. Engineers learn how op-amps can be employed to design active filters that offer precise control over frequency response characteristics. From high-pass and low-pass filters to band-pass and band-stop filters, the subchapter provides practical design guidelines, empowering engineers to tailor the filter response to meet specific requirements.

Furthermore, the subchapter discusses the utility of op-amps in analog computations and mathematical operations. Engineers are introduced to op-amp circuits that perform functions like addition, subtraction, integration, and differentiation. These circuits find applications in areas such as signal processing, control systems, and instrumentation, facilitating complex calculations with remarkable accuracy and speed.

Lastly, the subchapter explores op-amp-based voltage regulators and oscillators. Engineers discover how op-amps can be employed to regulate and stabilize voltages, ensuring a steady power supply for sensitive electronic components. Additionally, the subchapter delves into the principles of op-amp-based oscillators, revealing their significance in generating periodic waveforms for various applications, including clock signals, audio tones, and waveform synthesis.

In conclusion, this subchapter on operational amplifiers and their applications serves as a valuable resource for electrical engineers. By unraveling the mysteries behind op-amps and providing insights into their versatile applications, engineers gain the knowledge and skills necessary to leverage these powerful devices in their designs. Whether it is amplification, filtering, mathematical operations, voltage regulation, or waveform generation, op-amps continue to shape the landscape of electronics engineering, making them an indispensable tool for engineers in their quest for innovation and technological advancement.

Filters and Their Design Considerations

Filters play a crucial role in the field of electrical engineering, allowing engineers to manipulate signals and eliminate unwanted noise or frequencies. In this subchapter, we will delve into the various types of filters and the important design considerations that engineers must take into account when developing circuits.

First and foremost, it is essential to understand the different types of filters available. There are several common filter designs, including low-pass, high-pass, band-pass, and band-stop filters. Each filter type serves a specific purpose in signal processing and has its unique characteristics.

When designing a filter, engineers must carefully consider the desired cutoff frequency. This frequency determines the range at which the filter starts to attenuate the signal. The cutoff frequency is a critical parameter as it defines the boundary between the filtered and unfiltered regions. Determining the appropriate cutoff frequency requires a thorough understanding of the application and the specific signal requirements.

Another crucial consideration is the filter's roll-off rate or slope. The roll-off rate determines how quickly the filter attenuates frequencies beyond the cutoff point. A steeper roll-off rate is often desirable to eliminate unwanted frequencies effectively. However, a higher roll-off rate can also introduce phase distortion or signal loss. Engineers must strike a balance between the desired roll-off rate and potential trade-offs in signal quality.

In addition to the cutoff frequency and roll-off rate, engineers must also consider the filter's passband and stopband characteristics. The passband represents the range of frequencies that the filter allows to pass through without significant attenuation, while the stopband represents the range of frequencies that the filter attenuates. Careful selection of passband and stopband specifications is crucial to ensure the desired signal is retained while suppressing unwanted frequencies.

Moreover, engineers must consider the filter's impedance characteristics and its effect on the overall circuit. Impedance matching is crucial to prevent signal reflections and to ensure optimal power transfer. Understanding the impedance requirements of the filter and its interaction with the connected components is vital for designing a robust and efficient circuit.

Lastly, engineers must consider the practical constraints of the filter design, such as component availability, cost, and size. These factors can significantly impact the selection of filter topologies, component values, and implementation techniques.

In conclusion, filters are an indispensable tool for electrical engineers in signal processing and noise reduction. Designing effective filters requires careful consideration of parameters such as cutoff frequency, roll-off rate, passband and stopband characteristics, impedance matching, and practical constraints. By understanding these design considerations, engineers can develop high-quality filters that meet the specific requirements of their applications, ultimately leading to superior circuit performance and enhanced signal processing capabilities.

Chapter 5: Power Electronics and Power Management

Overview of Power Electronics

Power Electronics is a crucial field within electrical engineering that deals with the efficient conversion and control of electrical power. It encompasses the study and application of electronic devices and circuits for managing the flow of power in various systems, such as power supplies, motor drives, renewable energy systems, and electric vehicles.

In this subchapter, we will provide a comprehensive overview of power electronics, covering its fundamental principles, key components, and applications. Whether you are a seasoned electrical engineer or a novice in the field, this overview will serve as a valuable resource for understanding the essentials of power electronics.

To begin with, power electronics relies on the use of solid-state devices, such as power transistors, diodes, thyristors, and integrated circuits, to efficiently control and convert electrical energy. These devices enable the manipulation of voltage, current, and frequency, allowing power to be efficiently transferred between different sources and loads.

The subchapter will delve into the various types of power electronic converters, including rectifiers, inverters, and choppers. Each converter type serves a specific purpose, such as converting AC to DC, DC to AC, or changing the voltage levels. We will explore the operating principles, topologies, and control strategies employed in these converters, highlighting their advantages, limitations, and typical applications.

Furthermore, the subchapter will discuss the importance of power semiconductor devices in power electronics. We will delve into the characteristics, ratings, and selection criteria for different types of power devices, such as MOSFETs, IGBTs, and thyristors. Understanding these devices' capabilities and limitations is crucial for designing efficient and reliable power electronic systems.

Additionally, the subchapter will touch upon advanced topics in power electronics, such as soft switching techniques, resonant converters, and multilevel converters. These topics are of particular interest for engineers working on high-power and high-frequency applications, where efficiency and reduced electromagnetic interference are critical considerations.

In summary, this subchapter provides a comprehensive overview of power electronics, essential for electrical engineers working in various industries. By understanding the fundamental principles, key components, and applications of power electronics, engineers can design efficient and robust electrical systems, contributing to advancements in renewable energy, transportation, industrial automation, and more.

Power Semiconductor Devices and their Characteristics

In the ever-evolving field of electronics, power semiconductor devices have emerged as crucial components in various applications. From energy conversion to motor control and power management, these devices play a vital role in ensuring efficient and reliable operation. This subchapter aims to provide engineers, specifically electrical engineers, with a comprehensive understanding of power semiconductor devices and their characteristics.

To comprehend the significance of power semiconductor devices, it is essential to first grasp their basic structure and working principles. These devices, such as power diodes, bipolar junction transistors (BJTs), and metal-oxide-semiconductor field-effect transistors (MOSFETs), are designed to handle high current and voltage levels. They are characterized by their ability to switch between conducting and non-conducting states, thereby enabling efficient power conversion.

One of the key characteristics of power semiconductor devices is their ability to withstand high voltage levels. Engineers must understand the maximum voltage ratings of these devices to ensure their proper application and prevent failures. Moreover, they must consider the voltage drop across these devices and the impact on overall system efficiency.

Another crucial aspect is the ability of power semiconductor devices to handle high currents. Engineers need to be aware of the current ratings and thermal limitations of these devices to prevent overheating and subsequent failures. Proper heat sinking and thermal management techniques should be employed to ensure optimal device performance.

Switching characteristics are also crucial when working with power semiconductor devices. Engineers must comprehend the device's turn-on and turn-off times, as well as their switching losses. These characteristics directly affect the efficiency and reliability of power electronic systems.

Furthermore, engineers should be familiar with the concept of conduction losses in power semiconductor devices. These losses contribute to the power dissipation and require careful consideration during the design phase. Techniques such as pulse-width modulation (PWM) and zero-voltage switching (ZVS) can be utilized to minimize conduction losses and improve overall system efficiency.

Lastly, engineers must be aware of the various types of power semiconductor devices available in the market and their unique characteristics. This subchapter will provide an overview of different devices, including their advantages, disadvantages, and typical applications.

In conclusion, power semiconductor devices are essential components in modern electronic systems, particularly in the field of power electronics. Understanding their characteristics is crucial for electrical engineers to design efficient and reliable circuits. This subchapter will equip engineers with the necessary knowledge to select, apply, and optimize power semiconductor devices in their designs, ultimately contributing to the advancement of circuit design techniques.

Power Management Techniques and Circuit Design Considerations

In the field of electronics engineering, power management techniques and circuit design considerations play a crucial role in ensuring efficient and reliable operation of electronic devices. As electrical engineers, it is imperative to understand and implement these techniques to optimize power consumption, enhance performance, and extend battery life in various applications.

This subchapter aims to provide engineers, particularly electrical engineers, with a comprehensive understanding of power management techniques and circuit design considerations. It explores the fundamental principles, methodologies, and best practices that enable engineers to design circuits that are both energy-efficient and high-performing.

The subchapter begins by introducing the concept of power management and its significance in modern electronic systems. It emphasizes the importance of efficient power conversion, regulation, and distribution to meet the power demands of various electronic components. Furthermore, it delves into power loss mechanisms in circuits and discusses techniques to minimize power losses, such as utilizing low-power components, optimizing voltage levels, and employing efficient power supply topologies.

The subchapter also covers circuit design considerations that are essential for power management. It discusses the impact of component selection, layout design, and thermal management on power efficiency. It emphasizes the need for proper decoupling and filtering techniques to minimize noise and ensure stable power delivery.

Additionally, it explores the use of advanced design tools and simulation techniques to optimize power consumption and performance during the design phase.

Furthermore, the subchapter explores emerging power management techniques, such as energy harvesting, power factor correction, and dynamic power management. It explains how these techniques enable engineers to harness and utilize energy from alternative sources, improve power quality, and dynamically adjust power consumption based on system requirements.

To provide practical insights, the subchapter includes real-world examples, case studies, and design considerations for specific application domains. It covers topics such as power management in portable devices, automotive electronics, renewable energy systems, and IoT devices, addressing the unique challenges and considerations associated with each.

Overall, this subchapter serves as an invaluable resource for electrical engineers seeking to enhance their knowledge and skills in power management techniques and circuit design considerations. By mastering these principles and implementing efficient power management strategies, engineers can contribute to the development of energy-efficient and sustainable electronic systems.

Chapter 6: Radio Frequency (RF) Circuit Design

Introduction to RF Circuit Design

RF (Radio Frequency) circuit design is a crucial aspect of electrical engineering that deals with the design and implementation of circuits operating in the radio frequency range. This subchapter aims to provide engineers, specifically electrical engineers, with a comprehensive overview of RF circuit design techniques and principles.

In today's technologically advanced world, RF circuits play a vital role in various applications, including wireless communication systems, radar systems, satellite communication, and many more. Understanding the fundamentals of RF circuit design is essential for engineers to develop efficient and reliable RF systems.

The subchapter begins by introducing the basic concepts of RF circuit design, including an overview of the radio frequency spectrum and its applications. It then delves into the characteristics and challenges associated with RF circuits, such as impedance matching, noise, and interference. Engineers will learn how to analyze and optimize these parameters to ensure optimal performance.

Next, the subchapter explores the various components commonly used in RF circuit design, such as amplifiers, filters, oscillators, and mixers. It discusses the specific requirements and considerations when designing these components for RF applications, including frequency response, gain, stability, and linearity.

One crucial aspect of RF circuit design is the consideration of transmission lines and their effects on signal integrity. Engineers will learn about different types of transmission lines, including microstrip and stripline, and how to design and analyze them for minimal signal loss and distortion.

Furthermore, the subchapter covers RF circuit layout and grounding techniques, highlighting the importance of minimizing parasitic effects and maintaining signal integrity. It also introduces the use of simulation tools and software to aid in the design and analysis of RF circuits.

Throughout the subchapter, practical examples, case studies, and real-world applications are provided to illustrate the concepts discussed. This enables engineers to gain a deeper understanding of how RF circuits are applied in various industries.

By the end of this subchapter, electrical engineers will have acquired a solid foundation in RF circuit design. They will be equipped with the knowledge and skills necessary to tackle complex RF circuit design challenges, enabling them to develop cutting-edge RF systems that meet the demands of modern technology.

Overall, "Introduction to RF Circuit Design" serves as an essential resource for engineers seeking to expand their expertise in the field of electrical engineering, specifically within the niche of RF circuit design.

RF Amplifiers and Their Design Considerations

In the world of electronics engineering, amplifiers play a crucial role in enhancing the strength of signals, especially in the realm of radio frequency (RF) applications. RF amplifiers are specifically designed to boost weak signals and provide sufficient power for transmission, reception, and processing of RF signals. This subchapter will delve into the realm of RF amplifiers, discussing their design considerations and providing valuable insights for electrical engineers working in this niche.

One of the primary design considerations for RF amplifiers is gain. Gain refers to the amplification factor of the amplifier, which determines how much the input signal is amplified. Engineers need to carefully select the appropriate gain level based on the specific requirements of the application. It is essential to strike a balance between achieving sufficient amplification without introducing distortion or noise to the signal.

Another important aspect of RF amplifier design is bandwidth. To effectively process RF signals, amplifiers must be capable of handling a wide range of frequencies. Electrical engineers must consider the desired bandwidth and select components and design techniques that can accommodate the required frequency range. Additionally, stability is crucial when dealing with RF amplifiers, as any instability can lead to oscillations or unwanted feedback.

Efficiency is a key concern in RF amplifier design, as it directly impacts power consumption and heat dissipation. Engineers must aim to optimize the efficiency of the amplifier by selecting appropriate components, such as transistors and matching networks, and employing efficient biasing techniques.

Impedance matching is another critical consideration in RF amplifier design. It ensures that the input and output impedances of the amplifier are compatible with the source and load impedances, respectively. Proper impedance matching helps minimize signal reflection, maximize power transfer, and improve overall system performance.

Moreover, the linearity of an RF amplifier is crucial for accurately reproducing the input signal at the output without introducing distortion. Engineers must carefully design the amplifier circuit to minimize non-linearities and intermodulation distortions, ensuring clean and undistorted signal amplification.

Thermal management is yet another vital aspect of RF amplifier design. As RF amplifiers tend to generate significant heat, engineers must incorporate adequate cooling mechanisms to prevent overheating and maintain optimal performance.

In conclusion, this subchapter has explored the design considerations of RF amplifiers, providing valuable insights for electrical engineers working in the niche of RF electronics. By carefully considering factors such as gain, bandwidth, stability, efficiency, impedance matching, linearity, and thermal management, engineers can design RF amplifiers that meet the specific requirements of their applications while ensuring optimal performance and signal integrity.

RF Filters and Their Design Techniques

In the ever-evolving world of electronics engineering, the demand for high-performance RF filters has been skyrocketing. These filters play a crucial role in modern communication systems, ensuring optimal signal integrity and preventing unwanted interference. This subchapter aims to shed light on the fundamentals of RF filters and provide engineers, particularly electrical engineers, with a comprehensive understanding of their design techniques.

The subchapter begins with an introduction to RF filters, explaining their significance in various applications such as wireless communication, radar systems, and satellite communications. It emphasizes the need for effective filtering to achieve reliable and efficient signal transmission in these systems.

Next, the subchapter delves into the different types of RF filters commonly used in practice. It explores the characteristics, advantages, and limitations of various filter topologies, including low-pass, high-pass, band-pass, and band-stop filters. Real-world examples and case studies are included to illustrate the practical applications of these filters.

The subsequent sections focus on the design techniques for RF filters. The subchapter provides a step-by-step guide to designing filters with desired frequency response, bandwidth, and attenuation characteristics. It discusses the importance of understanding filter specifications, such as stopband ripple, passband insertion loss, and group delay, in achieving optimal filter performance.

Moreover, the subchapter introduces engineers to advanced design techniques, such as impedance matching, filter synthesis, and network analysis. It explores the use of advanced software tools and simulation techniques to aid in the filter design process, enabling engineers to save time and resources while improving filter performance.

To further enhance the engineer's understanding, the subchapter also covers the latest trends and advancements in RF filter technology. It discusses emerging design techniques, such as surface acoustic wave (SAW) filters and microelectromechanical systems (MEMS) filters, which offer improved performance and miniaturization possibilities.

In conclusion, "RF Filters and Their Design Techniques" provides electrical engineers with a comprehensive guide to designing high-performance RF filters. By familiarizing themselves with the principles, types, and design techniques discussed in this subchapter, engineers can confidently tackle the challenges of designing RF filters for various applications, contributing to the advancement of modern communication systems.

Chapter 7: Mixed-Signal Circuit Design

Introduction to Mixed-Signal Circuit Design

In today's technological landscape, mixed-signal circuit design holds a crucial position in the field of electrical engineering. As technology advances and the need for integrated circuits with both analog and digital functionalities becomes increasingly prevalent, engineers must possess a comprehensive understanding of mixed-signal circuit design principles. This subchapter aims to provide a concise yet insightful introduction to this fascinating field, catering specifically to electrical engineers.

Mixed-signal circuit design involves the integration of analog and digital circuits on a single chip, enabling the seamless interaction between the physical world and the digital domain. Unlike traditional analog or digital circuits, mixed-signal circuits bridge the gap between the two, allowing for the processing and manipulation of both continuous and discrete signals. This versatile approach has revolutionized various industries, including telecommunications, consumer electronics, and automotive systems.

The subchapter explores the fundamental concepts and methodologies that form the building blocks of mixed-signal circuit design. It begins with an overview of the key differences between analog and digital signals, highlighting their respective characteristics and challenges. Understanding the nuances of both signal types is essential for engineers to effectively design circuits that can handle the complexities of real-world applications.

Next, the subchapter delves into key components and techniques used in mixed-signal circuit design, such as amplifiers, filters, analog-to-digital converters (ADCs), and digital-to-analog converters (DACs). Each component is thoroughly explained, focusing on their roles, design considerations, and trade-offs. Furthermore, the subchapter explores various design strategies, including noise analysis, power management, and signal integrity, to ensure robust performance and optimize the overall system efficiency.

Additionally, the subchapter emphasizes the importance of simulation and modeling tools in mixed-signal circuit design. Engineers are introduced to industry-standard software and methodologies that enable accurate performance prediction and verification. Through simulation, engineers can analyze circuit behavior, optimize design parameters, and identify potential issues before fabrication, saving both time and resources.

Lastly, real-world examples and case studies are presented to illustrate how mixed-signal circuit design principles are applied in practical scenarios. These examples showcase the versatility and wide-ranging applications of mixed-signal circuits, inspiring engineers to explore new possibilities and push the boundaries of innovation.

In conclusion, this subchapter serves as a comprehensive introduction to mixed-signal circuit design, equipping electrical engineers with the foundational knowledge and tools necessary to tackle complex design challenges. By understanding and mastering this evolving field, engineers can contribute to the advancement of various industries and shape the future of technology.

Data Conversion Techniques

In the rapidly evolving field of electronics engineering, data conversion plays a critical role in enabling the seamless transmission and processing of information. As an electrical engineer, it is imperative to have a thorough understanding of various data conversion techniques to effectively design and optimize electronic circuits. This subchapter explores the fundamental concepts and advanced methods related to data conversion, providing engineers with valuable insights into this crucial aspect of circuit design.

The subchapter begins by introducing the concept of data conversion and its significance in modern electronic systems. It highlights the necessity of converting analog signals to digital format for processing, storage, and transmission purposes. The subchapter then delves into the principles of analog-to-digital conversion (ADC) and digital-to-analog conversion (DAC), shedding light on their underlying mechanisms and architectures.

Next, the subchapter discusses the different types of ADC and DAC circuits, emphasizing their advantages, limitations, and suitable applications. It covers successive approximation ADCs, flash ADCs, sigma-delta ADCs, and their respective working principles. Similarly, it explores different DAC architectures, including resistor string DACs, R-2R ladder DACs, and current-steering DACs, providing engineers with comprehensive knowledge of their operation and performance characteristics.

Furthermore, the subchapter delves into the key parameters and specifications associated with data converters, such as resolution, sampling rate, signal-to-noise ratio, and linearity. It explains how these parameters impact the overall performance and accuracy of data conversion, enabling engineers to make informed decisions during circuit design.

To enhance the engineers' understanding and practical implementation of data conversion techniques, the subchapter also includes case studies, simulation examples, and practical tips. It presents real-world scenarios where data conversion is crucial, such as audio and video processing, wireless communication systems, and sensor interfacing.

In conclusion, the subchapter on data conversion techniques equips electrical engineers with the knowledge and skills necessary to design and optimize electronic circuits. By understanding the principles, architectures, and parameters of ADCs and DACs, engineers can make informed decisions, ensuring accurate and efficient data conversion. This subchapter serves as an essential resource for engineers seeking to stay ahead in the dynamic field of electronics engineering and make significant contributions in their respective niches.

Design Considerations for Mixed-Signal Interfaces

In the field of electrical engineering, the integration of analog and digital circuits has become increasingly common. As a result, engineers are faced with the challenge of designing robust mixed-signal interfaces that can effectively handle the complexities associated with both analog and digital signals. This subchapter aims to provide a comprehensive overview of the key design considerations for mixed-signal interfaces, equipping engineers with the necessary knowledge to tackle these challenges.

First and foremost, one must understand the differences between analog and digital signals. Analog signals are continuous and can take on any value within a given range, while digital signals are discrete and can only have specific values (usually represented as 0 or 1). The interface between these two domains requires careful consideration to ensure accurate and reliable signal conversion.

One critical design consideration is noise. Analog signals are susceptible to noise, which can degrade the signal quality and introduce errors. Therefore, engineers must employ various techniques such as shielding, grounding, and filtering to minimize noise interference. Additionally, proper layout design and component selection are crucial to reduce noise coupling between analog and digital sections of the circuit.

Another important consideration is the choice of analog-to-digital converters (ADCs) and digital-to-analog converters (DACs). These converters play a pivotal role in converting analog signals to digital and vice versa. Engineers must carefully select converters with appropriate resolution, sampling rate, and linearity to ensure accurate signal conversion. Furthermore, the choice of converter topology (e.g., successive approximation, delta-sigma) and the implementation of anti-aliasing and reconstruction filters are critical in achieving high-fidelity signal conversion.

The interface between the analog and digital sections of a mixed-signal circuit also requires careful attention. Engineers must design proper signal conditioning circuits that interface between the two domains, ensuring compatibility and minimizing signal distortion. Techniques such as level shifting, buffering, and impedance matching are commonly employed to address signal compatibility issues and minimize reflections or signal integrity problems.

Lastly, power supply considerations cannot be overlooked. Mixed-signal interfaces often require separate power domains for analog and digital circuits to prevent interference and crosstalk. Engineers must carefully design power distribution networks, paying attention to decoupling capacitors, power integrity, and noise isolation techniques.

In conclusion, the design considerations for mixed-signal interfaces are crucial to ensure the seamless integration of analog and digital circuits. By understanding the differences between analog and digital signals, mitigating noise, selecting appropriate converters, addressing signal compatibility issues, and designing proper power supply networks, engineers can successfully design robust and reliable mixed-signal interfaces.

Chapter 8: Advanced Simulation and Modeling Techniques

Overview of Circuit Simulation Methods

In the field of advanced circuit design techniques, engineers, especially electrical engineers, need to have a deep understanding of circuit simulation methods. These methods play a crucial role in designing and analyzing complex electronic circuits, allowing engineers to save time, resources, and effort by virtually testing the circuit's behavior before physically implementing it.

This subchapter provides an overview of various circuit simulation methods commonly used in the industry. It aims to equip engineers with the necessary knowledge to choose the most suitable simulation method for their specific design requirements. The content covers both traditional and modern simulation techniques, highlighting their features, advantages, and limitations.

The subchapter begins by introducing the concept of circuit simulation and its importance in the design process. It emphasizes the benefits of simulating circuits, such as identifying potential issues, optimizing performance, and reducing prototyping costs. It also addresses the challenges faced by electrical engineers in circuit design and how simulation methods can help overcome these challenges.

The content then delves into the different types of circuit simulation methods. It starts with traditional methods like the nodal analysis, mesh analysis, and transient analysis. These methods are well-established and widely used for analyzing linear circuits. The subchapter explains their underlying principles and provides step-by-step instructions on how to perform these analyses.

Next, the subchapter explores more advanced simulation techniques, such as the Monte Carlo simulation, harmonic balance analysis, and sensitivity analysis. These methods are particularly useful for analyzing nonlinear circuits, RF circuits, and circuits with parameter variations. The content explains the principles behind these techniques and demonstrates their applications through practical examples.

Furthermore, the subchapter discusses the advantages and limitations of each simulation method, helping engineers make informed decisions based on their specific design requirements. It also introduces software tools commonly used for circuit simulation, such as SPICE (Simulation Program with Integrated Circuit Emphasis), and provides guidance on their usage and capabilities.

By the end of this subchapter, engineers will have a comprehensive understanding of circuit simulation methods and their applications in advanced circuit design. They will be equipped with the necessary knowledge to select the most appropriate simulation techniques for their projects, enabling them to design and analyze complex electronic circuits efficiently and effectively.

Model Development and Verification

In the realm of electrical engineering, model development and verification play a crucial role in the design and development of advanced circuitry. These techniques are essential for engineers to ensure the functionality, reliability, and efficiency of electronic systems. This subchapter aims to provide engineers with a comprehensive understanding of the techniques involved in model development and verification.

The first step in model development is to identify the system or component that requires modeling. It could be a complex integrated circuit, a power supply module, or even a simple resistor. Once the target is determined, engineers need to gather relevant data, such as electrical specifications, performance requirements, and environmental conditions.

The next phase involves choosing an appropriate modeling technique. Various methods are available, including analytical models, simulation models, and empirical models. Engineers must carefully evaluate the advantages and limitations of each technique and select the one that best suits their specific project.

Following the selection of a modeling technique, engineers embark on the development process. This typically involves creating mathematical equations or algorithms that accurately represent the behavior of the system or component under consideration. Accuracy is paramount, as any deviations could lead to faulty designs and suboptimal performance.

Once the model is developed, engineers move on to the verification phase. This step is critical to ensure that the model accurately predicts the behavior of the actual system or component. Verification is often performed through simulation, where the model is tested against a set of predefined inputs and compared with the corresponding outputs obtained from real-world measurements. Engineers meticulously analyze the simulation results and refine the model if necessary, repeating the verification process until a satisfactory level of accuracy is achieved.

The importance of model development and verification cannot be overstated. Through these techniques, engineers gain deep insights into the behavior of complex electronic systems, enabling them to optimize designs, identify potential issues, and make informed decisions. Moreover, accurate models facilitate effective collaboration among team members, as they provide a common language for discussing and evaluating designs.

In conclusion, model development and verification are indispensable tools for electrical engineers to design and develop advanced circuitry. By employing these techniques, engineers can ensure the functionality, reliability, and efficiency of electronic systems, ultimately contributing to the advancement of technology and innovation in the field of electrical engineering.

In the realm of advanced circuit design techniques, model development and verification play a vital role. For electrical engineers, understanding how to develop accurate models and verify their performance is crucial in designing efficient and reliable electronic systems. This subchapter aims to provide engineers with insights into the intricacies of model development and verification, equipping them with essential knowledge to excel in their field.

Model development is the process of creating mathematical representations of electrical components or systems. These models capture the behavior, characteristics, and interactions of various elements, allowing engineers to simulate and analyze their performance before physical implementation. The subchapter delves into the different types of models used in circuit design, such as linear and nonlinear models, equivalent circuit models, and behavioral models. It emphasizes the importance of selecting the appropriate model based on the specific requirements of the project.

Verification is the subsequent step in the design process, ensuring that the developed models accurately represent the real-world behavior of the circuit. Engineers must thoroughly validate the models by comparing simulated results with measured or expected data. This subchapter explores various techniques for model verification, including parameter estimation, model fitting, and sensitivity analysis. It also discusses the significance of statistical analysis in assessing the reliability and robustness of the models.

Furthermore, the subchapter highlights the significance of model validation through experimental testing. Engineers must conduct experiments to validate their models under different operating conditions and parameter variations. This validation process helps gain confidence in the accuracy and reliability of the models, ensuring they can be effectively applied to real-world scenarios.

To assist engineers in their model development and verification endeavors, the subchapter also introduces various software tools and simulation techniques commonly used in circuit design. These tools enable engineers to simulate, analyze, and optimize circuit performance, ultimately leading to more efficient and cost-effective designs.

In conclusion, model development and verification are essential components of advanced circuit design techniques for electrical engineers. By understanding the intricacies of model development, engineers can accurately represent the behavior of electrical components and systems. Through thorough verification and validation processes, they can ensure the reliability and effectiveness of these models. This subchapter serves as a comprehensive guide, equipping engineers with the knowledge and tools necessary to excel in their field and design cutting-edge electronics systems.

Advanced Simulation Techniques for Circuit Optimization

In today's rapidly evolving world of electronics engineering, it is crucial for electrical engineers to stay updated with advanced simulation techniques for circuit optimization. As the demand for high-performance circuits increases, engineers must employ innovative methods to design, analyze, and optimize circuits to meet the ever-growing expectations of modern electronic devices.

This subchapter sheds light on the latest simulation techniques that engineers can employ to optimize circuit designs. By leveraging these advanced techniques, electrical engineers can enhance the performance, efficiency, and reliability of their circuits while reducing costs and development time.

One of the key simulation techniques discussed in this subchapter is Monte Carlo analysis. This powerful method allows engineers to account for the inherent variability in circuit components, such as resistors, capacitors, and transistors. By performing numerous simulations with component values randomly varied within specified ranges, engineers can evaluate the circuit's performance under different operating conditions. This helps identify the potential weaknesses and sensitivities of the design, enabling engineers to make informed decisions and optimize the circuit accordingly.

Furthermore, the subchapter delves into sensitivity analysis, which enables engineers to determine the effect of parameter variations on the circuit's performance. By quantifying the impact of changes in component values, engineers can identify critical parameters that significantly influence the circuit's behavior. This knowledge allows them to focus their optimization efforts on these parameters, leading to more efficient and robust circuit designs.

In addition to Monte Carlo and sensitivity analysis, the subchapter explores other advanced simulation techniques, including optimization algorithms and design space exploration. These techniques empower engineers to automate the circuit optimization process, considering multiple objectives such as power consumption, speed, and area. By efficiently exploring the design space, engineers can quickly identify optimal solutions that meet the desired specifications.

Ultimately, mastering these advanced simulation techniques equips electrical engineers with the tools to push the boundaries of circuit design and create cutting-edge electronic devices. By optimizing circuit performance, engineers can meet the increasing demands of the market, enhance product quality, and contribute to the advancement of technology.

In conclusion, the subchapter on Advanced Simulation Techniques for Circuit Optimization provides electrical engineers with invaluable insights into the latest methods for optimizing circuit designs. By harnessing the power of techniques such as Monte Carlo analysis, sensitivity analysis, optimization algorithms, and design space exploration, engineers can elevate their circuit designs to new heights, ensuring high-performance, efficiency, and reliability in today's fast-paced electronic industry.

In the realm of circuit design, engineers are constantly striving to achieve optimal performance, efficiency, and reliability. To meet these demands, the field of circuit optimization has evolved significantly, offering engineers a wide range of advanced simulation techniques to enhance their design processes. In this subchapter, we will explore some of the cutting-edge simulation techniques that are revolutionizing circuit optimization for electrical engineers.

One of the key simulation techniques employed in circuit optimization is Monte Carlo analysis. This method allows engineers to evaluate the performance of a circuit by considering various parameter variations, such as component tolerances or process variations. By running numerous simulations with randomly selected parameter values, engineers can obtain statistical data that helps them identify potential weaknesses or areas of improvement in their circuit designs. Monte Carlo analysis provides engineers with a deeper understanding of the effects of parameter variations, ultimately leading to more robust and reliable circuit designs.

Another powerful simulation technique is sensitivity analysis. Sensitivity analysis allows engineers to determine the impact of parameter variations on the performance of a circuit. By selectively modifying specific parameters, engineers can assess the sensitivity of the circuit to these changes. This technique guides engineers in identifying critical parameters that significantly influence the overall performance of the circuit. With this knowledge, engineers can focus on optimizing these parameters to achieve desired performance goals.

Furthermore, advanced optimization algorithms are becoming increasingly popular in circuit design. These algorithms utilize mathematical techniques, such as genetic algorithms or particle swarm optimization, to automatically search for the best possible circuit configuration that meets specified design objectives. Through iterative optimization loops, engineers can quickly explore a vast design space and obtain optimal circuit solutions. This technique significantly reduces the time and effort required to achieve optimal circuit performance.

In addition to these simulation techniques, the subchapter will also delve into other advanced methods, including noise analysis, AC analysis, and transient analysis. These techniques enable engineers to study the behavior of circuits under different operating conditions, ensuring their designs meet performance criteria across a range of scenarios.

In conclusion, advanced simulation techniques are revolutionizing circuit optimization for electrical engineers. Through Monte Carlo analysis, sensitivity analysis, optimization algorithms, and other cutting-edge methods, engineers can achieve optimal circuit performance, efficiency, and reliability. By leveraging these techniques, engineers can expedite the design process, minimize costly iterations, and ultimately deliver superior circuit designs.

Chapter 9: Design for Manufacturability and Reliability

Design for Manufacturability (DFM) Principles

In the world of electrical engineering, designing circuits that are not only functional but also easily manufacturable is crucial. The concept of Design for Manufacturability (DFM) encompasses a set of principles that guide engineers in creating circuit designs that can be efficiently and cost-effectively manufactured. By adhering to DFM principles, engineers can streamline the production process, reduce manufacturing costs, and improve the overall quality of electronic products.

One of the fundamental DFM principles is to simplify the circuit design whenever possible. This involves reducing the number of components, minimizing the complexity of the circuit layout, and eliminating unnecessary features or functions. A simplified design not only makes the manufacturing process more efficient but also enhances reliability and reduces the chances of defects or failures.

Another important DFM principle is to carefully choose standard components. Using readily available, off-the-shelf components instead of custom-made ones not only saves time and effort but also lowers manufacturing costs. Standard components are typically produced in large quantities, allowing for economies of scale and easier sourcing.

Creating a design that is tolerant to variations in manufacturing processes is also a key DFM principle. Engineers should take into consideration the inevitable variations that occur during the manufacturing process and design circuits that can accommodate these variations without affecting overall performance. This can be achieved through proper component selection, using robust design techniques, and implementing built-in testing and calibration mechanisms.

Designing for ease of assembly is another critical DFM principle. Engineers should ensure that the circuit layout and component placement facilitate efficient assembly, minimizing the risk of errors or damage during the manufacturing process. This involves optimizing the placement of components, providing clear assembly instructions, and utilizing automated assembly techniques whenever possible.

Lastly, DFM principles emphasize the importance of designing for reliability and manufacturability testability. Engineers should incorporate design features that allow for thorough testing and inspection of the circuit at various manufacturing stages. This includes designing for easy access to test points, implementing on-board diagnostics, and incorporating redundancy or fault-tolerant mechanisms.

By embracing the principles of Design for Manufacturability (DFM), electrical engineers can create circuit designs that are not only functional but also optimized for efficient and cost-effective manufacturing. These principles, including simplifying the design, using standard components, accounting for manufacturing variations, designing for ease of assembly, and ensuring testability and reliability, are essential for successful electronic product development. By applying DFM principles, engineers can achieve higher yields, reduce production costs, and deliver high-quality products to the market.

In the world of electrical engineering, the design and manufacturing processes go hand in hand. It is crucial for engineers to not only create innovative and functional designs but also ensure that these designs can be efficiently and cost-effectively manufactured. This is where the principles of Design for Manufacturability (DFM) come into play.

DFM principles encompass a set of guidelines and techniques that help engineers optimize the manufacturability of their designs. By considering DFM principles from the initial stages of the design process, engineers can minimize risks, reduce production costs, enhance product quality, and accelerate time to market.

One fundamental DFM principle is simplification. Engineers should strive to simplify the design as much as possible without compromising functionality. By eliminating unnecessary complexities, engineers can streamline the manufacturing process, reduce the number of components, and enhance overall product reliability.

Another essential DFM principle is standardization. Engineers should aim to use standardized components and materials whenever possible. Standardization allows for easier sourcing, reduces lead times, and simplifies assembly processes. By adhering to industry standards, engineers can also ensure compatibility with existing manufacturing equipment and processes.

Designing for scalability is another critical DFM principle. Engineers should consider potential future production volume increases when creating their designs. By designing with scalability in mind, engineers can avoid costly redesigns or manufacturing process changes down the line as demand grows.

DFM principles also emphasize the importance of considering the manufacturing process during the design phase. Engineers should collaborate closely with manufacturing teams to understand their capabilities, constraints, and preferences. By aligning the design with the manufacturing process, engineers can optimize the use of available technologies and materials, leading to more efficient and cost-effective production.

Furthermore, the principles of Design for Manufacturability stress the importance of robustness and reliability. Engineers should design products that can withstand the rigors of the manufacturing process, transportation, and everyday use. By incorporating design features like proper tolerancing, stress analysis, and adherence to design guidelines, engineers can ensure the longevity and durability of their products.

In conclusion, Design for Manufacturability (DFM) principles are essential for electrical engineers to create designs that are not only innovative and functional but also efficiently and cost-effectively manufacturable. By embracing DFM principles, engineers can streamline the manufacturing process, reduce costs, enhance product quality, and ultimately deliver superior products to the market.

Reliability Analysis and Design Techniques

In the realm of advanced circuit design, engineers are constantly faced with the challenge of creating electronics systems that not only perform optimally but also exhibit high reliability. Reliability, in this context, refers to the ability of a circuit or system to function consistently and predictably over an extended period, even under adverse conditions or stress. This subchapter delves into the crucial aspect of reliability analysis and design techniques that electrical engineers need to consider when developing cutting-edge electronic systems.

Reliability analysis begins with an assessment of potential failure modes and their corresponding probabilities. Engineers must identify and understand the weak points or vulnerabilities in the circuit design, components, and manufacturing processes. By conducting detailed failure mode and effects analysis (FMEA), engineers can proactively address potential issues and implement suitable design modifications or redundancies to enhance reliability. Additionally, techniques such as fault tree analysis (FTA) can help identify the root causes of failures and guide engineers in devising effective preventive measures.

To ensure reliability, engineers must consider factors such as component selection, stress analysis, and robustness in their design process. Careful consideration is given to choosing components with proven track records and specifications that align with the intended application. Stress analysis techniques, such as thermal and mechanical simulations, aid in understanding how the circuit will perform under various operating conditions and external influences, enabling engineers to optimize the design for reliability.

Design for testability (DFT) is another crucial aspect covered in this subchapter. Engineers learn how to incorporate built-in self-test (BIST) features, boundary scan, and other testability techniques into their designs. These techniques assist in detecting and isolating faults during manufacturing, thereby improving the overall reliability of the system.

Furthermore, reliability design techniques encompass the implementation of redundancy and fault tolerance mechanisms. Engineers explore redundancy options such as dual-power supplies, backup circuits, and fault-tolerant architectures. By incorporating redundancy, engineers can ensure that the system can continue to function even in the presence of component failures or unexpected events, thus increasing the reliability and availability of the electronics system.

In conclusion, reliability analysis and design techniques are critical for electrical engineers to create robust and dependable electronic systems. By conducting thorough reliability analysis, selecting appropriate components, considering stress factors, implementing DFT features, and incorporating redundancy, engineers can enhance the reliability and longevity of their designs. This subchapter equips engineers with the knowledge and techniques necessary to tackle the challenges of reliability in advanced circuit design, enabling them to deliver highly reliable and efficient electronic systems.

In the dynamic and ever-evolving field of electronics engineering, the ability to design reliable circuits is of utmost importance. As electrical engineers, it is crucial to understand the principles and methodologies behind reliability analysis and design techniques to ensure the durability, efficiency, and longevity of electronic systems.

Reliability analysis refers to the systematic evaluation of the probability of failure or malfunction of a circuit or system. By identifying potential failure modes and determining their causes, engineers can implement design strategies to enhance reliability and minimize risks. This subchapter aims to delve into the key concepts and tools used in reliability analysis and design techniques, providing engineers with a comprehensive understanding of this critical aspect of circuit design.

One of the fundamental aspects of reliability analysis is understanding the concept of failure mechanisms. Different components within a circuit can fail due to various reasons, such as temperature, voltage stress, aging, and environmental factors. By studying failure mechanisms, engineers can identify potential weak points and develop strategies to mitigate them, ensuring the overall reliability of the system.

Reliability prediction models and software tools play a vital role in evaluating the reliability of electronic systems. This subchapter will explore the commonly used reliability prediction models, such as MIL-HDBK-217 and Telcordia, and discuss how to utilize them effectively. Additionally, engineers will gain insights into the application of advanced software tools that aid in reliability analysis, such as fault tree analysis and failure mode and effects analysis (FMEA).

Design techniques for reliability improvement will also be extensively covered in this subchapter. Topics such as redundancy, fault tolerance, derating, and stress analysis will be explored in great detail, equipping engineers with the knowledge and skills necessary to design robust and reliable circuits.

Furthermore, this subchapter will address the importance of testing and validation in ensuring the reliability of electronic systems. Various testing techniques, such as accelerated life testing and environmental stress screening, will be introduced to help engineers identify potential issues before the product reaches the end-users.

In conclusion, reliability analysis and design techniques are essential components of circuit design for electrical engineers. By understanding and implementing these techniques, engineers can create electronic systems that are durable, efficient, and dependable. This subchapter serves as a comprehensive guide, providing the necessary tools and knowledge to enhance the reliability of electronic systems, ultimately contributing to the advancement of the field of electronics engineering.

Failure Analysis and Prevention Strategies

In the challenging world of circuit design, failures are an inevitable part of the process. However, electrical engineers can minimize these failures by employing comprehensive failure analysis and prevention strategies. This subchapter explores the techniques and methodologies that can help engineers identify and rectify potential issues in circuit design, ensuring optimal performance and reliability.

Failure analysis is a systematic approach to understanding the root causes of circuit failures. By investigating the failure mechanisms, engineers can gain valuable insights into design flaws, manufacturing defects, or operational issues. This analysis involves a step-by-step examination of the failed components, conducting tests, and utilizing advanced tools such as scanning electron microscopes and thermal imaging cameras.

One of the primary objectives of failure analysis is to determine the failure mode and its underlying causes. This includes identifying factors such as overvoltage, overstress, temperature, mechanical stress, or material degradation. By understanding these factors, engineers can make informed decisions to improve circuit design and ensure its longevity.

Prevention strategies play a crucial role in avoiding circuit failures altogether. A proactive approach to design involves incorporating robust design practices and adhering to industry standards. Engineers must consider factors such as component selection, thermal management, electromagnetic interference (EMI) mitigation, and reliability testing. By integrating these considerations into the design process, engineers can reduce the risk of failures and enhance the overall quality of the circuit.

Another vital aspect of prevention strategies is the implementation of comprehensive testing and validation procedures. Rigorous testing at various stages of the design cycle, including simulations, prototype testing, and environmental stress testing, helps identify potential issues early on. This allows engineers to make necessary modifications before mass production, ensuring that the circuit meets the desired performance and reliability requirements.

Furthermore, failure analysis data can be utilized to improve future designs. By closely examining past failures, engineers can identify patterns, common failure modes, or weaknesses in design methodologies. This knowledge can guide the development of more robust and reliable circuits, preventing similar failures in subsequent designs.

In conclusion, failure analysis and prevention strategies are essential for electrical engineers to ensure the reliability and performance of circuit designs. By analyzing failures, understanding their causes, and implementing preventive measures, engineers can minimize the risks associated with circuit failures. This subchapter provides valuable insights and practical techniques that will empower engineers to create superior circuit designs, contributing to the advancement of the field of electronics engineering.

In the field of electrical engineering, the design and implementation of electronic circuits is a complex and critical task. Engineers are constantly faced with the challenge of ensuring the reliability and functionality of their designs. However, despite their best efforts, failures can still occur, leading to costly setbacks and potential safety hazards. This subchapter aims to explore failure analysis and prevention strategies, equipping electrical engineers with the knowledge and tools to mitigate the risks associated with circuit design.

Failure analysis is a systematic approach to identify the root causes of failures in electronic circuits. It involves investigating various factors, such as environmental conditions, material properties, and manufacturing processes, to determine the underlying mechanisms that lead to failure. By understanding the specific reasons behind failures, engineers can develop effective prevention strategies.

One of the key prevention strategies is thorough testing and simulation. Engineers must conduct rigorous testing at different stages of the design process to identify any potential weaknesses or vulnerabilities in the circuit. This can be achieved through simulation software that can accurately model the behavior of the circuit under various conditions. By simulating different scenarios, engineers can identify potential failure points and make necessary design modifications to improve reliability.

Another important aspect of failure prevention is component selection. Engineers must carefully evaluate the quality and reliability of the components they choose for their circuits. Using subpar or counterfeit components can significantly increase the risk of failure. It is crucial to work with reputable suppliers and conduct thorough research to ensure the components meet the required specifications and standards.

Additionally, engineers should consider environmental factors when designing circuits. Factors such as temperature, humidity, and vibration can significantly impact the performance and longevity of electronic circuits. By taking into account these environmental conditions during the design process, engineers can implement appropriate measures, such as proper thermal management and protective coatings, to enhance the circuit's reliability.

Furthermore, failure analysis can provide valuable insights into the design and manufacturing processes. By analyzing failure patterns and trends, engineers can identify areas for improvement and implement preventive measures. This includes enhancing quality control procedures, refining manufacturing techniques, and improving documentation and communication within the design team.

In conclusion, failure analysis and prevention strategies are crucial for electrical engineers to ensure the reliability and functionality of electronic circuits. By conducting thorough failure analysis, engineers can identify the root causes of failures and develop effective prevention strategies. Through rigorous testing, careful component selection, consideration of environmental factors, and continuous process improvement, engineers can significantly reduce the risk of failures and enhance the overall reliability of their circuit designs.

Chapter 10: Emerging Trends in Circuit Design

Introduction to Emerging Technologies

In this rapidly advancing era of technology, it is crucial for electrical engineers to stay updated with the latest trends and breakthroughs in their field. The world is witnessing the emergence of various cutting-edge technologies that are revolutionizing the way we live, work, and interact. This subchapter aims to provide engineers, specifically electrical engineers, with an introduction to these emerging technologies and their potential applications.

The field of electronics engineering is constantly evolving, with new technologies being developed and integrated into various industries. These emerging technologies hold immense potential for transforming the way electrical engineers design circuits and solve complex problems. By understanding and harnessing the power of these technologies, engineers can stay ahead of the curve and create innovative solutions to meet the demands of the future.

One of the most exciting emerging technologies is the Internet of Things (IoT), which refers to the interconnectivity of everyday objects through the internet. IoT has the potential to revolutionize various industries such as healthcare, transportation, and energy management. Electrical engineers play a crucial role in developing devices and systems that enable seamless communication and data exchange between these interconnected objects.

Another emerging technology that has gained significant attention is Artificial Intelligence (AI) and Machine Learning (ML). AI and ML algorithms can analyze large amounts of data and make intelligent decisions, leading to automation and optimization of various processes. Electrical engineers can utilize AI and ML techniques to improve the design and performance of electronic circuits, leading to more efficient and intelligent systems.

Furthermore, advancements in renewable energy technologies, such as solar and wind power, are transforming the way we generate and consume electricity. Electrical engineers are at the forefront of designing and implementing innovative solutions to harness and store renewable energy efficiently.

In addition to these examples, this subchapter will also explore other emerging technologies, including augmented reality (AR), virtual reality (VR), nanotechnology, and quantum computing. Each of these technologies has the potential to disrupt traditional approaches to circuit design and open up new possibilities for engineers.

As electrical engineers, it is essential to embrace and adapt to these emerging technologies. By doing so, engineers can enhance their skills, expand their knowledge, and unlock new opportunities for innovation. This subchapter will serve as a stepping stone for electrical engineers to explore and incorporate these emerging technologies into their design practices, enabling them to stay at the forefront of their field and drive technological advancements.

In today's rapidly evolving world, it is essential for electrical engineers to stay up-to-date with the latest advancements in technology. The field of electronics is constantly witnessing the emergence of new and innovative technologies that have the potential to revolutionize the way we design and develop circuits. This subchapter aims to provide engineers, specifically electrical engineers, with an introduction to these emerging technologies and their potential applications.

Advancements in technology have led to the development of various emerging technologies that are reshaping the landscape of circuit design. One such technology is the Internet of Things (IoT), which refers to the interconnection of devices through the internet, enabling them to communicate and exchange data. IoT has opened up new possibilities for electrical engineers, allowing them to create smart and interconnected systems that can monitor and control various aspects of our daily lives.

Another significant emerging technology is artificial intelligence (AI) and machine learning (ML). These technologies have gained immense popularity due to their ability to process large amounts of data and make intelligent decisions. Electrical engineers can leverage AI and ML to design circuits that can adapt and learn from their environment, paving the way for autonomous systems and intelligent devices.

Furthermore, the rise of renewable energy sources has propelled the development of technologies such as solar power, wind energy, and energy storage systems. Electrical engineers play a crucial role in designing circuits that efficiently capture and convert renewable energy into usable forms, contributing to a sustainable future.

Additionally, advancements in nanotechnology have opened up new possibilities for circuit design. Nano-electronics offer improved performance, miniaturization, and energy efficiency. Engineers can utilize nanotechnology to design circuits with faster speeds, reduced power consumption, and increased integration.

Moreover, the subchapter will explore emerging technologies such as quantum computing, 5G wireless communication, and wearable electronics. These technologies have the potential to revolutionize various industries and create new opportunities for electrical engineers.

In conclusion, staying updated with emerging technologies is crucial for electrical engineers to ensure they remain at the forefront of circuit design. This subchapter provides a comprehensive introduction to various emerging technologies, including IoT, AI, renewable energy, nanotechnology, quantum computing, 5G, and wearable electronics. By understanding these technologies, engineers can leverage their potential to design innovative circuits that will shape the future of electronics.

Circuit Design Considerations for IoT Devices

In today's interconnected world, the Internet of Things (IoT) has become an integral part of our daily lives. From smart homes to wearable devices, IoT has revolutionized the way we interact with technology. However, designing circuits for IoT devices presents unique challenges that require careful consideration. This subchapter will delve into the various circuit design considerations specifically tailored for electrical engineers working in the IoT niche.

One of the primary concerns when designing circuits for IoT devices is power consumption. IoT devices are often battery-powered, and therefore, optimizing power usage is crucial to ensure long battery life. Engineers must carefully select components that have low power consumption and design circuits that minimize energy usage during both active and standby modes. Techniques such as power gating, clock gating, and voltage scaling can be employed to achieve power efficiency without compromising performance.

Another crucial consideration is the size and form factor of the IoT device. Many IoT applications require small, compact devices that can fit seamlessly into various environments. Electrical engineers must carefully select components and design circuits that are not only efficient but also compact. Incorporating integrated circuits, surface-mount technology, and miniaturized components can help achieve the desired form factor for IoT devices.

Furthermore, IoT devices often rely on wireless communication protocols such as Bluetooth, Zigbee, or Wi-Fi to connect to the internet. Engineers must understand the intricacies of these protocols and design circuits that ensure reliable and secure wireless communication. Signal integrity, interference mitigation, and power amplification are among the key considerations when designing the communication circuitry for IoT devices.

Security is another critical aspect of IoT device design. As these devices collect and transmit sensitive data, it is imperative to implement robust security measures to protect against unauthorized access and data breaches. Engineers must incorporate encryption algorithms, secure authentication mechanisms, and tamper-resistant components into the circuit design to ensure the utmost security for IoT devices.

Lastly, engineers must consider the manufacturability and cost-effectiveness of the circuit design. Choosing readily available components, designing for ease of assembly, and optimizing the manufacturing process can help reduce production costs and facilitate mass production of IoT devices.

In conclusion, designing circuits for IoT devices requires engineers to address specific considerations such as power consumption, form factor, wireless communication, security, manufacturability, and cost-effectiveness. By carefully considering these factors, electrical engineers can ensure the successful development of efficient, compact, secure, and commercially viable IoT devices that seamlessly integrate into our interconnected world.

In today's interconnected world, the Internet of Things (IoT) has revolutionized the way we interact with technology. The ability to connect various devices and enable seamless communication has opened up a plethora of opportunities for engineers, particularly electrical engineers, to design innovative IoT devices. However, designing circuits for IoT devices comes with its own set of challenges and considerations. This subchapter explores the important circuit design considerations for engineers working on IoT devices.

Power consumption is a critical factor in IoT devices due to their wireless nature and the need for long battery life. Engineers must carefully choose low-power components and optimize the circuit design to minimize power consumption. Techniques like power gating, clock gating, and voltage scaling can be employed to achieve power efficiency without compromising performance.

Another crucial consideration is the size and form factor of the IoT device. As IoT devices are often embedded within everyday objects, their size and shape should be compact and unobtrusive. Engineers need to design circuit boards that are small and lightweight without sacrificing functionality. Advanced packaging techniques such as System-in-Package (SiP) and Chip-on-Board (CoB) can be employed to achieve miniaturization.

Wireless connectivity is a fundamental aspect of IoT devices, enabling seamless communication and data exchange. Engineers must carefully select appropriate wireless protocols, such as Wi-Fi, Bluetooth, or Zigbee, based on the specific requirements of the IoT application. The circuit design should incorporate the necessary components, such as antennas and transceivers, to ensure reliable wireless connectivity.

Security is a paramount concern in IoT devices, as they often handle sensitive data. Engineers should implement robust security measures at the circuit design level, including encryption, secure authentication, and tamper detection mechanisms. Additionally, the circuit design should incorporate secure boot mechanisms to ensure the integrity of the device's firmware.

Finally, interoperability and compatibility with other IoT devices and platforms should be considered during circuit design. Engineers should adhere to industry standards and protocols to ensure seamless integration and communication with other devices, cloud platforms, and IoT ecosystems.

In conclusion, designing circuits for IoT devices requires electrical engineers to consider various factors such as power consumption, size, wireless connectivity, security, and interoperability. By carefully addressing these considerations, engineers can create efficient, reliable, and secure IoT devices that seamlessly integrate into our interconnected world.

Future Directions in Advanced Circuit Design Techniques

In the ever-evolving field of electronics engineering, staying abreast of the latest advancements in circuit design techniques is crucial for engineers to remain at the cutting edge of innovation. As technology continues to push the boundaries of what is possible, it is essential to explore and anticipate the future directions of advanced circuit design techniques. This subchapter delves into the exciting prospects and emerging trends that electrical engineers can expect to encounter in the near future.

One of the key areas that holds great promise is the integration of artificial intelligence (AI) in circuit design. With the advent of machine learning algorithms and neural networks, engineers can leverage AI tools to optimize circuit performance, reduce power consumption, and enhance overall system reliability. By leveraging AI, engineers can automate the labor-intensive process of circuit design, allowing for faster and more efficient development cycles.

Furthermore, the future of advanced circuit design techniques lies in the exploration of novel materials and components. As traditional silicon-based technologies reach their physical limits, engineers are beginning to explore alternatives such as graphene, carbon nanotubes, and organic semiconductors. These materials offer unique properties that can revolutionize circuit design, enabling higher speeds, increased power efficiency, and flexibility.

Another exciting direction in advanced circuit design techniques is the integration of wireless communication capabilities into circuits. With the rise of the Internet of Things (IoT) and the increasing demand for interconnected devices, engineers must develop circuits that can seamlessly communicate with each other wirelessly. This integration opens up new possibilities for smart homes, wearable devices, and autonomous systems, creating a future where everything is interconnected.

In addition to these advancements, engineers can look forward to the development of energy-efficient circuit design techniques. As sustainability becomes a global priority, electrical engineers are tasked with designing circuits that minimize power consumption and reduce environmental impact. This involves exploring energy harvesting techniques, efficient power management strategies, and the utilization of renewable energy sources.

In conclusion, the future of advanced circuit design techniques for electrical engineers is exciting and full of potential. By embracing AI, exploring novel materials, integrating wireless communication capabilities, and prioritizing energy efficiency, engineers can shape a future where technology seamlessly integrates into our lives, empowering us to achieve greater heights of innovation. As the field continues to evolve, it is imperative for engineers to stay curious, adaptable, and open to new possibilities to drive progress and shape the future of electronics engineering.

As technology continues to evolve at an unprecedented pace, the field of advanced circuit design techniques for electrical engineers is also undergoing significant advancements. This subchapter explores the future directions that engineers in this niche can expect to see in the coming years.

One of the key areas that will shape the future of circuit design is the integration of artificial intelligence (AI) and machine learning (ML) algorithms. These technologies have the potential to revolutionize circuit design by enabling engineers to create more optimized and efficient circuits. AI and ML algorithms can analyze vast amounts of data, identify patterns, and generate innovative solutions that were previously unimaginable. With the incorporation of AI and ML in circuit design, engineers can expect to see faster and more accurate designs, reduced power consumption, and enhanced performance.

Another significant trend in advanced circuit design is the increasing focus on low-power design techniques. With the growing demand for portable and battery-operated devices, engineers are now challenged with designing circuits that consume minimal power while maintaining optimal performance. In the future, innovative low-power design techniques, such as voltage scaling, clock gating, and power gating, will play a crucial role in circuit design. These techniques will not only extend battery life but also contribute to reducing the overall environmental impact of electronic devices.

The emergence of Internet of Things (IoT) technologies is another area that will heavily influence advanced circuit design techniques. With the proliferation of IoT devices, engineers will need to design circuits that are capable of handling massive amounts of data, ensuring robust connectivity, and maintaining security. Additionally, the miniaturization of IoT devices will require engineers to develop innovative packaging and integration techniques to optimize space utilization and ensure reliable performance.

Furthermore, the adoption of advanced semiconductor technologies, such as nanoscale integrated circuits, will revolutionize circuit design. These technologies offer unprecedented levels of integration, speed, and power efficiency. Engineers will need to adapt their design techniques to leverage the benefits of nanoscale integrated circuits while addressing the challenges associated with manufacturing, reliability, and thermal management.

In conclusion, the future of advanced circuit design techniques for electrical engineers holds immense potential and exciting possibilities. The integration of AI and ML algorithms, focus on low-power design, IoT technologies, and advanced semiconductor technologies will shape the way circuits are designed. As engineers explore these future directions, they will be able to create innovative and efficient circuit designs that cater to the evolving needs of the electronics industry.

Chapter 11: Case Studies and Practical Applications

Case Study 1: Designing a High-Speed Digital Circuit

Introduction:

In the fast-paced world of electronics engineering, designing high-speed digital circuits is a challenging but essential task. These circuits form the backbone of modern electronic devices, ranging from smartphones and laptops to advanced medical equipment and aerospace systems. This case study delves into the intricacies of designing a high-speed digital circuit, shedding light on the critical considerations and techniques that engineers must employ to ensure optimal performance.

Understanding High-Speed Digital Circuits:

High-speed digital circuits operate at frequencies in the range of hundreds of megahertz to several gigahertz. Unlike traditional low-speed circuits, they demand meticulous attention to signal integrity, noise management, and power integrity. The increased clock rates and data transfer rates in these circuits introduce various challenges that require careful analysis and design.

Design Considerations:

1. Signal Integrity: Maintaining signal integrity is crucial to avoid data corruption and ensure reliable transmission. Engineers must consider factors like impedance matching, controlled trace lengths, and proper termination techniques. The study explores the impact of signal

reflections, crosstalk, and parasitic capacitance on signal integrity, providing valuable insights into mitigating their effects.

2. Noise Management: High-speed digital circuits are vulnerable to noise sources, such as electromagnetic interference (EMI) and power supply noise. Employing proper grounding techniques, shielding, and noise filtering mechanisms is essential to minimize noise-induced errors and disruptions. The case study highlights effective noise management strategies to enhance circuit performance.

3. Power Integrity: A key challenge in high-speed digital circuit design is managing power distribution effectively. Engineers must analyze power delivery networks (PDNs) to ensure stable voltage levels and minimize voltage drops. Techniques like decoupling capacitors, power plane design, and optimized routing are explored in this case study to address power integrity concerns.

4. Timing Analysis: Accurate timing analysis is imperative in high-speed digital circuits to avoid issues like data skews, setup/hold violations, and clock jitter. The study presents methodologies for analyzing timing margins, ensuring reliable operation even under worst-case scenarios.

Conclusion:

Designing high-speed digital circuits demands a deep understanding of signal integrity, noise management, power integrity, and timing analysis. By leveraging advanced techniques and careful considerations, engineers can overcome the challenges posed by these circuits and achieve high-performance designs. This case study serves as a valuable resource for electrical engineers seeking to enhance their expertise in designing high-speed digital circuits, enabling them to contribute to the development of cutting-edge electronic devices across various industries.

Introduction:

In the rapidly evolving field of electronics engineering, the design of high-speed digital circuits holds immense significance. This subchapter presents a comprehensive case study that delves into the intricacies of designing a high-speed digital circuit. By exploring the challenges, considerations, and solutions involved in this process, engineers can enhance their understanding and expertise in this specialized niche of electrical engineering.

Understanding High-Speed Digital Circuits:

High-speed digital circuits refer to electronic systems that operate at remarkably high frequencies, often surpassing several gigahertz. These circuits are commonly found in applications such as telecommunications, computer networking, and digital signal processing. Designing a high-speed digital circuit demands meticulous planning, precise timing, and the careful management of various signal integrity issues.

Challenges and Considerations:

When dealing with high-speed digital circuits, engineers encounter several challenges that can significantly impact circuit performance. These challenges include signal integrity degradation, noise, crosstalk, electromagnetic interference (EMI), and power consumption. Additionally, designing for high-speed circuits requires a deep understanding of transmission line theory, impedance matching, and the selection of appropriate components.

Solutions and Techniques:
This case study presents a step-by-step approach to designing a high-speed digital circuit, focusing on key techniques and solutions to overcome the aforementioned challenges. It covers topics such as layout considerations, power distribution network design, choosing suitable transmission lines, and reducing signal reflections. Furthermore, the case study explores the importance of using proper decoupling capacitors, grounding techniques, and shielding to mitigate noise and eliminate EMI issues.

Simulation and Testing:
To ensure the robustness and reliability of the designed high-speed digital circuit, engineers must incorporate simulation and testing methodologies. This subchapter discusses the utilization of advanced simulation tools, such as SPICE and electromagnetic field solvers, to verify the circuit's performance before fabrication. Additionally, it emphasizes the significance of conducting various tests, including eye diagrams, timing analysis, and jitter measurements, to validate the circuit's functionality and meet the desired specifications.

Conclusion:
Designing a high-speed digital circuit requires a comprehensive understanding of signal integrity, transmission line theory, and noise management techniques. By studying this case study, electrical engineers can enhance their knowledge and skills in this specialized niche. With a focus on overcoming challenges and implementing effective solutions, this subchapter equips engineers with the necessary tools to design robust and high-performing high-speed digital circuits.

Case Study 2: Designing an Analog Filter

Analog filters play a crucial role in various electrical engineering applications, including signal processing and communication systems, where the need to eliminate unwanted frequencies and enhance desired ones is essential. In this case study, we will explore the process of designing an analog filter, highlighting the key considerations and techniques employed by electrical engineers.

1. Understanding the Design Requirements:
Before embarking on the filter design journey, it is crucial to have a clear understanding of the project's specifications. Engineers must identify the desired frequency response, such as the cutoff frequency, passband ripple, and stopband attenuation. Additionally, other requirements, including input/output impedance, power consumption, and size constraints, must be taken into account.

2. Selecting the Filter Type:
Once the design requirements are established, engineers need to choose the appropriate filter type that best suits the application. Common filter types include low-pass, high-pass, band-pass, and band-stop. Each type has its specific characteristics and advantages, which need to be considered in light of the design requirements.

3. Designing the Filter Circuit:
After selecting the filter type, engineers proceed to design the filter circuit. This involves selecting the appropriate passive components, such as resistors, capacitors, and inductors, and determining their values. Furthermore, active components, like operational amplifiers, might be required to achieve the desired filter characteristics.

Engineers need to consider component tolerances, temperature coefficients, and other non-ideal effects during this phase.

4. Simulation and Analysis:
To ensure the effectiveness of the designed analog filter, engineers employ simulation tools to model and analyze the circuit's behavior. Through simulations, engineers can validate whether the filter meets the desired specifications, identify potential issues, and fine-tune the design parameters.

5. Prototyping and Testing:
Once the design is simulated and refined, engineers move on to prototyping the analog filter circuit. This involves physically assembling the circuit and conducting various tests to validate its performance. Engineers rely on specialized test equipment, such as signal generators, oscilloscopes, and spectrum analyzers, to measure the filter's frequency response, gain, and other parameters.

6. Optimization and Iteration:
The prototyping and testing phase often reveals areas for improvement. Engineers analyze the test results, identify any discrepancies, and optimize the design accordingly. This iterative process continues until the filter meets all the design requirements and performs optimally.

In conclusion, designing analog filters requires a systematic approach and a deep understanding of the design requirements. Electrical engineers must carefully select the appropriate filter type, design the circuit using passive and active components, simulate and analyze the circuit's behavior, prototype and test the design, and iterate for optimization. By following these steps, engineers can create analog filters that meet the stringent demands of various electrical engineering applications.

Introduction:
In this case study, we will delve into the intricacies of designing an analog filter, a fundamental component in electronic circuits. Analog filters play a crucial role in signal processing, eliminating unwanted noise and improving the overall performance of electrical systems. As electrical engineers, understanding the design principles and techniques behind analog filters is essential for creating robust and efficient circuitry.

Designing the Filter:
To begin the design process, engineers must first determine the specific requirements of the filter. Factors such as the desired frequency response, gain, and filter type (e.g., low-pass, high-pass, band-pass) must be carefully considered. Once these parameters are established, engineers can choose the appropriate filter topology and design methodology.

Selection of Filter Topology:
Various filter topologies exist, each with its own advantages and limitations. Some commonly used topologies include Butterworth, Chebyshev, and Bessel filters. These topologies differ in terms of their frequency response characteristics, such as passband ripple, stopband attenuation, and phase response. Engineers must select the topology that best suits the requirements of their application.

Design Methodology:
After selecting the filter topology, engineers can proceed to the design phase. This involves selecting the component values, such as resistors, capacitors, and inductors, that determine the filter's characteristics. Design equations and software tools can aid in calculating these values accurately. Engineers must also consider practical constraints, such as component tolerances and availability.

Simulation and Analysis:
Once the filter design is complete, engineers employ simulation software to verify its performance. Simulations help identify potential issues and allow for iterative refinement of the design. Parameters such as gain, phase shift, and frequency response can be analyzed to ensure they meet the desired specifications.

Prototyping and Testing:
After successful simulation, engineers proceed to the prototyping stage. They construct a physical circuit based on the design and conduct thorough testing. This involves applying input signals of varying frequencies and amplitudes to the filter and analyzing the output. Engineers must verify that the filter meets the desired specifications and make any necessary adjustments.

Conclusion:
Designing an analog filter requires a holistic approach, encompassing theoretical knowledge, practical considerations, and simulation tools. By carefully selecting the appropriate filter topology, designing the circuit, simulating its performance, and testing the physical prototype, engineers can develop robust analog filters that meet the requirements of their applications. Understanding the intricacies of analog filter

design is crucial for electrical engineers to ensure optimal signal processing and enhance overall circuit performance.

Case Study 3: Designing a Power Electronics System

Introduction:
In this subchapter, we will explore a real-life case study that focuses on the design of a power electronics system. Power electronics plays a crucial role in various electrical engineering applications, ranging from renewable energy systems to electric vehicles. Through this case study, engineers will gain valuable insights into the design considerations, challenges, and best practices involved in developing efficient and reliable power electronics systems.

Understanding the Problem:
The case study revolves around the development of a power electronics system for a solar energy plant. The objective is to design an efficient system that can convert the DC power generated by the solar panels into AC power suitable for grid integration. This involves selecting the appropriate power semiconductor devices, designing the power conversion stages, implementing control algorithms, and ensuring the system's overall stability and reliability.

Design Considerations:
To accomplish the goal of designing a power electronics system, engineers need to consider several key factors. These include the system's efficiency, power density, thermal management, electromagnetic compatibility (EMC), and cost-effectiveness. Each aspect requires a careful trade-off analysis to achieve an optimal solution.

Challenges and Solutions:
Power electronics system design poses various challenges, such as

managing high power levels, minimizing losses, mitigating electromagnetic interference, and ensuring safe operation. Engineers must address these challenges by employing advanced techniques like soft switching, optimal control strategies, and advanced cooling techniques. The case study will present specific examples and solutions to overcome these challenges.

Best Practices:
Throughout the case study, engineers will learn about best practices for power electronics system design. These include proper component selection, thermal management techniques, efficient power conversion topologies, and robust control algorithms. The case study will also emphasize the importance of simulation and prototyping to validate and optimize the system's performance before implementation.

Conclusion:
This case study provides valuable insights into the design process of a power electronics system for a solar energy plant. By exploring the challenges, considerations, and best practices involved, electrical engineers can enhance their understanding and skills in developing efficient and reliable power electronics systems. This knowledge will enable them to tackle similar projects successfully and contribute to the advancement of renewable energy and other electrical engineering applications.

Introduction:
In this subchapter, we will delve into a fascinating case study that focuses on designing a power electronics system. Power electronics is an essential field within electrical engineering that deals with the efficient conversion, control, and transfer of electrical power from one

form to another. This case study will provide engineers with a comprehensive understanding of the design process involved in creating a power electronics system, highlighting key considerations, challenges, and innovative solutions.

Designing the Power Electronics System: The case study begins by outlining the specific requirements and objectives of the power electronics system, such as power conversion efficiency, voltage regulation, and load stability. We will explore the different topologies and architectures available for power electronics systems and discuss the selection process based on the application's needs. Emphasis will be placed on system reliability, cost-effectiveness, and scalability.

Components Selection and Integration: Next, we will examine the critical components required for the power electronics system, such as power converters, semiconductor devices, capacitors, and inductors. Engineers will gain insights into the selection criteria for each component, considering factors such as voltage and current ratings, switching speed, thermal management, and EMI/EMC considerations. Additionally, we will explore the integration challenges and techniques for optimizing the system's performance.

Control and Protection Mechanisms: Designing a robust control and protection system is crucial for ensuring the safe and efficient operation of the power electronics system. This case study will delve into the various control strategies available, including pulse width modulation (PWM) techniques, feedback loops, and digital control algorithms. We will also discuss the implementation of protection mechanisms such as overvoltage, overcurrent, and overtemperature protection circuits.

Testing, Simulation, and Validation:
To ensure the reliability and functionality of the power electronics system, engineers must conduct thorough testing, simulation, and validation. We will explore various simulation tools, such as SPICE and MATLAB/Simulink, to analyze the system's performance under different operating conditions. Additionally, practical testing methods and techniques will be discussed, including prototype development, measurement, and analysis.

Conclusion:
In conclusion, this case study on designing a power electronics system provides engineers with a comprehensive understanding of the principles, methodologies, and challenges involved in creating efficient and reliable power conversion solutions. By exploring the various stages of the design process, component selection, control mechanisms, and testing methodologies, engineers will be equipped with the necessary knowledge to tackle real-world power electronics design projects effectively. This case study serves as a valuable resource for electrical engineers seeking to enhance their expertise in power electronics and advance their careers in this dynamic field.

Chapter 12: Conclusion and Future Perspectives

Summary of Key Concepts

In this subchapter, we will provide a concise summary of the key concepts covered in the book "Advanced Circuit Design Techniques for Electronics Engineers." This summary is specifically tailored to engineers, particularly those in the niche of electrical engineering. By understanding these key concepts, you will be equipped with the necessary knowledge and techniques to design complex circuits efficiently and effectively.

1. Signal Integrity: Signal integrity is crucial in circuit design as it ensures the quality of transmitted signals. This concept explores techniques to minimize noise, distortion, and interference, enabling reliable data transmission.

2. High-Speed Design: With the increasing demand for high-speed electronic devices, engineers must have a deep understanding of high-speed design techniques. This includes managing signal propagation delays, controlling impedance, and reducing crosstalk.

3. Power Integrity: Power integrity focuses on maintaining a stable and clean power supply to the circuit components. It involves minimizing voltage drops, managing power distribution networks, and reducing electromagnetic interference.

4. Electromagnetic Compatibility (EMC): EMC is vital to ensure that electronic devices operate without interfering with each other and comply with regulatory standards. This concept covers shielding,

grounding, and filtering techniques to minimize electromagnetic emissions and susceptibility.

5. Analog and Mixed-Signal Design: Analog and mixed-signal circuits play a crucial role in many applications. This section explores techniques for designing amplifiers, filters, and analog-to-digital converters, ensuring accurate signal processing and conversion.

6. RF and Microwave Design: RF and microwave circuits are essential for wireless communication systems. This concept covers topics such as impedance matching, S-parameters, microwave filters, and amplifiers.

7. Printed Circuit Board (PCB) Design: PCBs are the backbone of electronic systems, and their design greatly impacts the overall circuit performance. This section discusses layout considerations, stack-up design, and routing techniques for optimal signal integrity and thermal management.

8. Design for Manufacturability (DFM): DFM focuses on designing circuits that are easy to manufacture, test, and assemble. This concept explores various design guidelines to minimize manufacturing costs, improve yield, and enhance reliability.

By grasping these key concepts, electrical engineers can enhance their circuit design skills, overcome challenges, and design innovative and efficient electronic systems. The book "Advanced Circuit Design Techniques for Electronics Engineers" provides in-depth explanations, practical examples, and case studies to further expand your knowledge in these areas.

In this subchapter, we will provide a concise overview of the key concepts covered in the book "Advanced Circuit Design Techniques for Electronics Engineers." This summary is specifically tailored to address electrical engineers who are seeking to deepen their understanding of circuit design and enhance their skills in this field.

1. Integrated Circuit Design: We delve into the intricacies of integrated circuit design, exploring essential topics such as transistor-level design, layout techniques, and design optimization. By understanding these concepts, electrical engineers can develop more efficient and high-performance integrated circuits.

2. Signal Integrity: Maintaining signal integrity is crucial for ensuring reliable and error-free communication within electronic systems. We discuss various techniques and strategies to mitigate signal integrity issues, including signal routing, transmission line theory, and impedance matching.

3. Power Electronics: Power electronics play a vital role in modern electronic systems, particularly in power conversion and management. Our book covers advanced power electronics concepts, including switching devices, topologies, and control techniques, empowering engineers to design efficient and reliable power systems.

4. Analog Circuit Design: Analog circuits are fundamental to many electronic systems, and we provide an in-depth exploration of advanced analog circuit design techniques. Topics covered include operational amplifiers, filters, voltage references, and noise analysis, enabling engineers to design robust analog circuits.

5. RF Circuit Design: Radio frequency (RF) circuits are essential for wireless communication systems, and we offer an overview of RF circuit design principles and techniques. This section covers topics such as impedance matching, RF filters, oscillators, and mixers, equipping engineers with the knowledge to design high-frequency circuits.

6. Design for Testability: Testing electronic circuits is crucial to ensure their proper functioning and reliability. We discuss various design considerations and techniques to enhance testability, including built-in self-test, fault diagnosis, and boundary scan techniques.

Throughout the book, practical examples, case studies, and design tips are provided to illustrate the application of these concepts in real-world scenarios. We aim to empower electrical engineers with the knowledge and skills required to tackle complex circuit design challenges and achieve optimal performance in their designs.

Whether you are a seasoned electrical engineer or a recent graduate looking to expand your knowledge, "Advanced Circuit Design Techniques for Electronics Engineers" offers a comprehensive guide to mastering advanced circuit design concepts and techniques.

The Importance of Continuous Learning in Circuit Design

In the ever-evolving field of circuit design, continuous learning is not just a choice but a necessity for electrical engineers. With technology advancing at an unprecedented pace, it is crucial for engineers to stay up-to-date with the latest advancements, techniques, and tools in circuit design. This subchapter explores the importance of continuous learning in circuit design and highlights the benefits it brings to electrical engineers.

First and foremost, continuous learning allows engineers to stay ahead of the curve in this competitive industry. By keeping abreast of the latest trends and developments, engineers can design circuits that are not only efficient and reliable but also innovative. With new components, software, and methodologies emerging regularly, engineers who invest in continuous learning gain a competitive advantage over their peers and can offer cutting-edge solutions to complex design challenges.

Furthermore, continuous learning helps engineers enhance their problem-solving skills. As they delve deeper into new concepts and techniques, engineers gain a broader perspective on circuit design principles. This enables them to approach design problems from different angles and develop creative solutions. By continuously expanding their knowledge base, engineers can tackle complex design issues with confidence, ultimately improving the quality and efficiency of their work.

Another significant benefit of continuous learning in circuit design is the ability to adapt to the ever-changing industry requirements. The demand for energy-efficient and compact electronic devices is skyrocketing, pushing engineers to find innovative ways to optimize circuit designs. By learning about emerging technologies such as Internet of Things (IoT), artificial intelligence, and renewable energy sources, engineers can adapt their designs to meet these new challenges. Continuous learning equips electrical engineers with the skills and knowledge necessary to design circuits that are tailor-made for the future.

Lastly, continuous learning fosters professional growth and personal fulfillment. As engineers expand their knowledge and expertise, they become valuable assets to their organizations and the industry as a whole. Moreover, learning new skills and mastering advanced techniques brings a sense of accomplishment and satisfaction to engineers, fueling their passion for circuit design and motivating them to push the boundaries of what is possible.

In conclusion, continuous learning is vital for electrical engineers in the field of circuit design. By staying current with the latest advancements, engineers can remain competitive, enhance their problem-solving skills, adapt to industry requirements, and experience professional growth. Embracing continuous learning is not just a professional responsibility but an opportunity to excel and make a meaningful impact in the world of circuit design.

In the fast-paced world of electronics engineering, staying up-to-date with the latest advancements and techniques is crucial for success. Circuit design, being the backbone of any electronic system, requires engineers to continuously learn and adapt to new methodologies and technologies. This subchapter explores the importance of continuous learning in circuit design and its impact on the work of electrical engineers.

Technology is evolving at an unprecedented rate, with new components and integrated circuits being introduced regularly. To stay ahead of the competition, engineers need to constantly upgrade their knowledge and skills. Continuous learning allows electrical engineers to keep pace with these advancements and gain a deeper understanding of the intricacies involved in circuit design.

One of the key reasons why continuous learning is vital in circuit design is the need for engineers to optimize designs for performance, efficiency, and reliability. By staying updated with the latest design tools and methodologies, engineers can ensure that their circuits are efficient, consume less power, and have minimal error rates. Continuous learning enables engineers to implement cutting-edge techniques such as high-speed signal integrity, noise reduction, and power management, resulting in superior designs.

Moreover, continuous learning fosters innovation and creativity in circuit design. As engineers gain new knowledge and skills, they are better equipped to think outside the box and develop novel solutions to complex design challenges. This not only enhances their problem-solving abilities but also allows them to push the boundaries of what is possible in circuit design.

Another crucial aspect of continuous learning in circuit design is the ability to troubleshoot and resolve issues effectively. As engineers learn about the latest diagnostic tools and techniques, they can quickly identify and rectify any problems that may arise during the design process. This reduces development time, minimizes costs, and ensures that the final product meets the desired specifications.

In conclusion, continuous learning is of utmost importance in circuit design for electrical engineers. It enables them to stay updated with the latest advancements, optimize designs for performance and efficiency, foster innovation, and effectively troubleshoot issues. By embracing continuous learning, engineers can ensure that their circuit designs are at the forefront of technology, meeting the ever-increasing demands of the electronic industry.

Future Perspectives in Advanced Circuit Design Techniques

As the field of electronics continues to evolve, engineers are constantly seeking new and innovative circuit design techniques to meet the growing demands of the industry. The future of advanced circuit design holds tremendous potential, offering exciting opportunities for electrical engineers to push the boundaries of what is possible.

One of the key areas that will shape the future of circuit design is the development of smaller, more efficient and power-saving devices. As technology progresses, there is an increasing need for circuits that can operate with minimal power consumption while still delivering high performance. Engineers will need to focus on developing techniques that optimize power usage, reduce heat dissipation, and improve overall energy efficiency.

Another important aspect of future circuit design is the integration of advanced materials and technologies. As new materials with unique properties become available, engineers will have the opportunity to explore novel circuit design approaches. For example, the use of nanomaterials and graphene promises to revolutionize circuit design by enabling faster switching speeds, higher frequencies, and improved device performance. Additionally, the incorporation of flexible and stretchable electronics opens up new possibilities for wearable devices and smart textiles.

The future of circuit design also lies in the realm of artificial intelligence (AI) and machine learning (ML). With the increasing complexity of circuits, engineers will need AI and ML algorithms to assist in the design process. These intelligent algorithms can help optimize circuit performance, identify potential issues, and suggest improvements, ultimately leading to more efficient and reliable designs.

Furthermore, the rise of the Internet of Things (IoT) will drive the need for circuit designs that can seamlessly integrate with a vast array of connected devices. Engineers will need to develop techniques that enable efficient communication, data processing, and power management within IoT systems. This will require a deep understanding of wireless communication protocols, sensor integration, and low-power circuit design.

In conclusion, the future of advanced circuit design techniques holds immense potential for electrical engineers. From power-saving designs and advanced materials to AI-assisted design processes and IoT integration, engineers must stay at the forefront of technological advancements to meet the demands of the ever-evolving electronics industry. By embracing these future perspectives, engineers can drive innovation and shape the future of circuit design.

As electrical engineers, it is crucial to stay abreast of the latest advancements in circuit design techniques to meet the ever-increasing demands of the industry. In this subchapter, we explore the future perspectives in advanced circuit design techniques that are set to revolutionize the field.

One of the key areas that hold immense potential is the integration of artificial intelligence (AI) in circuit design. AI can significantly enhance the efficiency and accuracy of the design process by automating various tasks, such as optimization, layout, and verification. With the help of machine learning algorithms, engineers can quickly explore a vast design space and identify optimal solutions, saving time and effort. Furthermore, AI can also aid in the development of intelligent circuits that can adapt and self-reconfigure based on real-time conditions, leading to more robust and flexible electronic systems.

Another promising trend is the emergence of flexible and stretchable electronics. As the demand for wearable devices and conformable electronics continues to grow, engineers need to develop circuit design techniques that can accommodate these unique requirements. Flexible electronics rely on novel materials and manufacturing processes, where circuit designs must be able to withstand bending, stretching, and even twisting. By embracing these challenges, engineers can unlock new opportunities for applications in healthcare, consumer electronics, and beyond.

Moreover, the rapid progress in nanotechnology opens up exciting possibilities for advanced circuit design techniques. Nano-scale devices, such as carbon nanotubes and graphene, exhibit exceptional electrical properties, making them ideal for high-performance circuits. However, designing circuits at the nano-scale poses numerous challenges, including manufacturing precision, reliability, and power management. By developing innovative design methodologies tailored for nanoscale devices, engineers can harness their full potential and pave the way for ultra-miniaturized and energy-efficient electronic systems.

Additionally, the Internet of Things (IoT) revolution is driving the need for low-power and energy-efficient circuit design techniques.

With billions of connected devices expected to be deployed in the coming years, power consumption becomes a critical concern. Engineers must focus on developing power-aware design methodologies that optimize energy consumption without compromising performance. Techniques such as power gating, voltage scaling, and dynamic power management will play a pivotal role in meeting the energy efficiency requirements of IoT devices.

In conclusion, the future of advanced circuit design techniques is incredibly promising and offers engineers a multitude of opportunities. By embracing AI, flexible electronics, nanotechnology, and energy-efficient designs, engineers can push the boundaries of innovation and create electronic systems that are smarter, more adaptable, and environmentally friendly. As electrical engineers, it is our responsibility to continually explore and adopt these future perspectives to contribute to the advancement of the field.

Printed in the USA
CPSIA information can be obtained
at www.ICGtesting.com
LVHW021219020524
778879LV00015BA/967

9 788119 747498